Philosophy
of Science

FUNDAMENTALS OF PHILOSOPHY SERIES
Series Editors
John Martin Fischer, University of California, Riverside
John Perry, Stanford University and University of California, Riverside

MIND: A BRIEF INTRODUCTION
John R. Searle

BIOMEDICAL ETHICS
Walter Glannon

A CONTEMPORARY INTRODUCTION TO FREE WILL
Robert Kane

POLITICAL PHILOSOPHY
A. John Simmons

DEBATING SEX AND GENDER
Georgie Warnke

PHILOSOPHY OF SCIENCE: A NEW INTRODUCTION
Gillian Barker, Philip Kitcher

Philosophy of Science

A NEW INTRODUCTION

GILLIAN BARKER
Western University

PHILIP KITCHER
Columbia University

NEW YORK   OXFORD
OXFORD UNIVERSITY PRESS

Oxford University Press is a department of the University of Oxford.
It furthers the University's objective of excellence in research,
scholarship, and education by publishing worldwide.

Oxford New York
Auckland Cape Town Dar es Salaam Hong Kong Karachi
Kuala Lumpur Madrid Melbourne Mexico City Nairobi
New Delhi Shanghai Taipei Toronto

With offices in
Argentina Austria Brazil Chile Czech Republic France Greece
Guatemala Hungary Italy Japan Poland Portugal Singapore
South Korea Switzerland Thailand Turkey Ukraine Vietnam

Copyright © 2014 by Oxford University Press

Published by Oxford University Press
198 Madison Avenue, New York, New York 10016
http://www.oup.com

Oxford is a registered trademark of Oxford University Press

Library of Congress Cataloging-in-Publication Data
Barker, Gillian.
  Philosophy of science : a new introduction / Gillian Barker, Western University, Philip Kitcher,
Columbia University.
      pages cm
Includes bibliographical references and index.

ISBN 978-0-19-536619-8
1. Science—Philosophy. 2. Philosophy and science. I. Title.
Q175.B217 2013
501—dc23
                                    2013000895
Printing number: 9 8 7 6 5 4 3 2 1

Printed in the United States of America on acid-free paper

To the memory of Carl Gustav (Peter) Hempel

In whose footsteps we all walk.

Contents

Preface

This book is about the philosophy of science. We know from experience that this expression sounds, to many people, almost like a contradiction in terms. What could philosophy and science have to do with one another? Philosophy seems preoccupied with profound problems that can never be resolved: the "eternal questions" of the meaning of life and the nature of knowledge and the good. Science seems precisely the opposite: cut and dried, simply concerned with concrete matters of fact. Yet science and philosophy have vitally important things to say to one another. The sciences have transformed—and continue to transform—our understanding of the world we live in and of our place in it, our history and our future; the new understanding they have given us has implications that can be felt through every branch of philosophy. On the other hand, closer scrutiny reveals that the sciences raise deep and pressing philosophical questions of their own. Scientific claims have tremendous authority in today's societies, and many of us believe that scientific inquiry is able to give us a special kind of knowledge: insight into the underlying workings of the natural world that is uniquely objective and reliable. Yet the sciences are also contested, subject to internal dispute among experts as well as to criticism from without. When public debates about any particular scientific issue become heated, the questions raised are philosophical ones about the nature, authority, and ownership of scientific knowledge. To make choices in our lives, we must each come to some conclusions about how to think about scientific controversies on issues as diverse as health risks and global climate change. At a political level, we face additional questions about how to shape public policy in response to the conflicting claims of scientists and of their critics, and about how to make choices about the direction of science itself. All of these questions require us to think philosophically about science. This book aims to show what such thinking looks like, and why it is both important and fascinating to do it.

There are many introductory books in the philosophy of science, and very high standards were set by C. G. Hempel's *Philosophy of Natural Science*, first published in 1966. For many subsequent authors, the task has seemed to be one of updating Hempel's book. We, too, see that book as a classic, but we think the time has come for a large expansion of the agenda. Developments in philosophy, in history of science, in sociology of science, in the sciences themselves, and in public reactions to them—all occurring in the past decades— have made it important to raise, and discuss, questions that Hempel and his followers put to one side. There are many books published since 1966 in which the topics of our Chapter 2 are pursued at greater depth—but, for us, that is one chapter among many.

We have had help from many other people in bringing this idea to realization. Kristin Maffei and Robert Miller have made the process as easy as anyone could hope for. Generous and perceptive reviewers helped greatly in shaping the book; we have tried to meet the challenges they set: Justin B. Biddle (Georgia Institute of Technology), Frank K. Fair (Sam Houston State University), Bruce Glymour (Kansas State University), Richard Grandy (Rice University), Michael Hoffmann (Georgia Institute of Technology), Marc Lange (UNC Chapel Hill), Nicholas Thompson (Clark University), and Michael Weisberg (University of Pennsylvania) all contributed signally. Gillian particularly thanks Jonathan Barker (for close reading and many good ideas), Nancy Barker (for unflagging support and many good dinners), and above all Dave Pearson (for everything). Philip particularly thanks Patricia Kitcher (as always an exemplary reader and the most constructive of critics).

Science and
Philosophy

Scientific Disputes and Philosophical Questions

For more than three decades now, researchers who investigate the Earth's climate have been telling the rest of the world that our planet is heating up, and that human activities are largely responsible. During the past two centuries people have released an increasing amount of carbon dioxide into the atmosphere, with the result that some of the heat that would normally dissipate is trapped. Many investigators believe that the long-term consequences for life on Earth are serious, and that the future well-being of our species is profoundly endangered. Yet although there have been periodic upsurges of interest in restricting the emission of greenhouse gases, and despite the efforts of a few nations to reduce dependence on fossil fuels, there is no coherent global strategy for responding to the supposed threat.

Why has science failed to carry the day? Why has debate about the scientific case continued? In many nations there are vocal groups who deny that the alleged facts have been established. Journalists and politicians talk of the myth of global warming; large conglomerates fund "alternative" research; apparently moderate voices point out that the specific consequences of whatever warming trend has been established are matters of dispute and that policies designed to limit emissions might plunge the world into an economic crisis having even more adverse effects on our descendants. So, it is supposed, the question should remain open.

The problem is that science alone cannot tell us how to make reasonable judgments about what is happening to the world's climates, or about how we should respond to the threat of climate change. To do this, we need good science, but we also have to face some basic questions that science can't fully answer: How exactly have the climatologists arrived at their conclusions?

What is the evidence and what does it entitle people to believe? How should we craft policies for the future when we recognize the uncertainty of our own situation? Whose interests should be taken into account and how should conflicting needs be weighed against one another? These are philosophical questions. They arise from an important problem that confronts humanity, and from the role science plays in our efforts to understand and address that problem.

Global warming is not an isolated case. Developments in the sciences often call for philosophical reflection. Consider another case, one in which scientific research is entangled with how we think about ourselves and other people. During much of recent human history people have categorized one another by race. Moreover, they have frequently operated with a view that some races are naturally—intrinsically—inferior to others. Sometimes research in the sciences has supported these claims. So, for instance, it has been asserted that intelligence is measured by scores on a particular test, that there are differences in the average scores of members of different races, and that studies of twins who have been reared apart reveal that intelligence is highly "heritable." Other scientists have disputed both the data and the interpretation offered by those who would defend deep racial differences. Some have suggested that a systematic study of the world's people reveals no basis for thinking that our species is divided into races, and that we should eliminate the concept of race entirely. Almost all would now agree that there is no evidence whatsoever for the existence of genes that have any noteworthy effect on cognitive abilities or traits of character and that are unequally distributed across the groups marked as "races." Yet recent research in molecular genetics does show that DNA sequences with no known import (bits of what is sometimes thought of as "junk DNA") are found with different frequencies in populations that have been isolated from one another for a significant period of time, so that there are "natural" divisions of *Homo sapiens* into smaller groups that share a closer kinship. Popular discussions of that research often view it as rehabilitating the notion of race.

Is that correct? What are we saying when we suppose that a particular division of the living world (or of the inorganic world) is "natural"? On what evidence are claims like this based? How should we explain the features of human psychology and behavior that fascinate us, and account for the differences across various populations?

Think about another pair of examples, not normally juxtaposed. Physicists have sometimes campaigned for public funds to build large facilities in which they hope to accelerate the weird microentities they view as the fundamental constituents of matter to speeds so high that their collisions would produce a type of particle that has been theoretically predicted but never detected. (American physicists lost in their attempt to secure government money, but their European counterparts won, and they appear now to have found their elusive target.) On a more modest scale, Freudian psychoanalysts advertise themselves as having a method, grounded in an understanding of the constituents and mechanisms of the mind, that enables them to bring relief to people with psychiatric troubles. Despite the increased popularity of drugs as remedies for psychiatric disorders, as well as the emergence of alternative forms of psychotherapy, some analysts continue to attract patients and to make a comfortable living.

In both instances, the entities that inspire various practical procedures—building huge tunnels, weekly sessions on the couch—are both strange and remote from everyday observation and detection. How can we fathom the mysteries of the Higgs boson or identify a repressed conflict between a patient and his father? Are these things part of reality, on a par with apples and oranges, rocks and radios? Or should we think of particle physics and psychoanalysis simply as practical devices, good insofar as they lead people to the goals they want to achieve, but not as making any serious claims about nature? Is there a significant difference between the two instances, and, if so, in what does it consist?

We could continue the list, but these few examples are probably enough to make the point. All over the map of contemporary science, further questions—nonscientific questions—arise. As you ponder those questions, you are led to issues that seem to lie in the province of philosophy. What is evidence, and how do we obtain it? How should people act when they can recognize that their evidence is partial? Does the world come with natural divisions, and, if so, how can they be discovered? Is it right to think of the sciences as giving a deep picture of nature, even when the things it discusses are strikingly at odds with our previous ideas about reality? Who has the authority to make scientific judgments, and why?

Overarching these questions are even more general ones. Are the natural sciences the uniquely best sources of human knowledge, setting standards that ought to be achieved in all fields of inquiry? Do they constitute just one of many ways of thinking about ourselves and the world that are good in different ways or that serve different purposes? Do they threaten our understanding of ourselves, presenting a limited or distorting vision of the world and our place in it?

The philosophy of science, as we understand it, consists in an attempt to answer—or, at least, discuss—these questions, both the more specific ones and those that are most general. This book is an introduction to it.

Modern Science: A Brief History

Let us start more slowly and more systematically. What are we talking about when we talk about "science"?

We start with examples. The natural sciences include physics, chemistry, biology, earth science, atmospheric science, oceanography, neuroscience, and at least some parts of psychology. We exclude mathematics because its ways of establishing its conclusions seem so different—most obviously, mathematicians don't appear to rely on observation and experimentation. We mostly leave out applied sciences (like metallurgy), engineering, and technology, although it is not always easy to separate these forms of inquiry from "pure" (or "basic") science. The social sciences (economics, anthropology, linguistics, etc.) pose some of the same issues as those that arise for the natural sciences, but also bring distinctive questions of their own that we do not pursue. As we shall see, defining what exactly makes a field of investigation count as a science is itself a task that raises substantial philosophical questions—these are among the main concerns of Chapter 2. It is clear nonetheless that the natural sciences on our list share features that make it reasonable to treat them together, and that they raise important and interesting common issues that philosophy can usefully address. The resulting investigation ultimately sheds some light on work in mathematics, the applied sciences, and the social sciences as well.

Current work in the natural sciences can be traced to inquiries that were pursued in ancient Greece. The word *science* is, however, a recent coinage, introduced with its present meaning only in the nineteenth century. Before that, the investigations that aimed at general and systematic explanations of nature were known as natural philosophy; more particular descriptive studies were parts of natural history. Even in the twentieth century, undergraduates at Cambridge University studying science were prepared for exams in natural philosophy. The enduring label testifies to an old intertwining of science and philosophy.

The practices of the sciences as we now know them have been profoundly shaped by the events of the seventeenth century. Galileo and Kepler, Bacon and Descartes, and Boyle and Newton were the most prominent figures in a transition that is often known as the Scientific Revolution (although some historians of science maintain that that label suggests a change more abrupt and complete than the reality). It is hard to deny that a fundamental shift occurred. To approach some of the questions we will be discussing, and to appreciate the appeal of particular answers to them, it is necessary to have some understanding of what happened.

At the beginning of the sixteenth century, most Europeans shared a picture of the cosmos drawn from the works of Aristotle and interpreted from a Christian perspective. According to this picture, the Earth was located at the center of the universe, and the moon, planets, sun, and stars revolved around it. Earth itself, and all objects found on and near its surface, were composed of four elements: earth, air, fire, and water. Each of these elements had different qualities that helped explain the properties of the objects they composed. Earthly objects were subject to change and decay. The stars, planets, and other objects in the heavens were composed of a special fifth substance, quintessence; they were eternal, flawless, and unchanging.

Events were explained in terms of causes, but among the causes that were deemed important for explanation and understanding were some of a special sort. Each kind of object was thought to have a "final cause" or *telos,* something like a goal or purpose, determining the changes and motions it naturally tends to undergo. Objects could be made to move unnaturally or "violently," but their natural motions were the basis of the regularities we observe and it was important to identify them in providing a general explanation. Earthly objects moved naturally in straight lines toward or away from the center of the universe, at the center of the Earth. The natural motion of heavenly bodies was circular, for (as Greek philosophers had supposed) the circle is the simplest and most perfect geometrical figure, and circular motion can continue eternally. Some heavenly bodies moved in simple circles around the center of the universe, but others were thought to move on paths compounded of various circular movements (this concession was made to account for the complex movements of the planets against the background of the "fixed stars"). The mathematical structure of the heavens had been worked out in detail by Ptolemy in the second century. Sometimes this geocentric universe was understood as a real physical structure of nested rotating crystalline spheres on which the various heavenly bodies were borne. Ptolemy's system required a great many compound movements (known as epicycles: see Figure 1.1), however, which made such a realistic interpretation difficult to maintain. The Ptolemaic system was often thought of as a mere calculating device for generating predictions, not as a representation of real objects and causes.

Inquiry into the natural world was guided by the writings of earlier scholars, in particular those of Aristotle and the theologians who had integrated his work with Christian doctrine. Investigation began with a careful exegesis of the views of these authorities, and sought to answer questions in

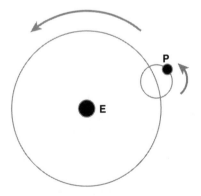

Figure 1.1 Epicycles in the Ptolemaic system.

ways that agreed with those views as properly understood. Empirical observations were recruited to assist such investigations, but qualitative observations made in everyday circumstances or reference to familiar observable phenomena were sufficient for the purposes of natural philosophy, and no special techniques or quantitative measurements were called for.

The sixteenth century began to challenge this picture of the world and the associated ideas about proper investigation. Trouble began quite accidentally. A Polish monk, Nicolaus Copernicus, was asked by the Papacy to develop a better model of the motions of the heavenly bodies—it was evident that the calendar needed reform, and improved astronomy was the key to solving the practical problem. Copernicus labored for more than thirty years, ultimately producing an astronomical system that displaced the Earth from the center of the universe. It was clear from the start that the new model was easier to work with than the traditional versions—and also evident that for any heliocentric system there was a geocentric system that would deliver exactly the same predictions for the planetary motions (although the geocentric models would be more cumbersome). Orthodox astronomy embraced Copernicus by treating the new system as a mere device for making astronomical calculations. That way embarrassing questions could be dodged: How did the birds and clouds keep up with a moving Earth? What exactly happened in the biblical event when Joshua commanded the sun to stand still? But a very few of Copernicus's successors wondered if the heterodox ideas of the new astronomy might be literally true—if perhaps, after all, the Earth did really move. Early in the seventeenth century, Galileo's observations of the moon and planets with the new telescope provided reason to think Copernicus had correctly represented the real structure of the universe.

Galileo's contemporary, Johann Kepler, another one of the rare Copernicans, struggled to find a more accurate representation of the motions of the planets, eventually replacing the perfect circles with elliptical orbits. After Galileo and Kepler, many thinkers came to view the Earth not as the center around which the universe was built, but merely as one planet among many. They also thought in novel ways about the causes of change and the nature of explanation. Aristotle's distinction between natural and violent motion was replaced with the idea that motion is the same everywhere—in Earth as it is in Heaven—and governed everywhere by the same natural laws. Eventually, toward the end of the seventeenth century, Isaac Newton's mathematical laws of motion and of gravitation revealed how the whole system worked.

At the same time, investigators in other areas were developing a new account of matter. Explanations of the properties of substances in terms of the four Aristotelian elements gave way to mechanistic theories that attributed the qualities of objects to mechanical properties (shape and motion) of their component parts, often conceived as indivisible atoms. The cosmos of the

Middle Ages, small and intimate, filled with purpose and value, centered around the uniquely important Earth with its human inhabitants, was replaced by a universe understood as analogous to a vast clock, with an order that was grand and comprehensible but wholly impersonal.

Investigators who turned to the authority of their own reason and senses challenged the reliance on tradition. Galileo and Bacon championed the importance of making observations and experiments. More radically, Descartes, troubled perhaps by the breakdown of Aristotelian ideas that had dominated inquiry for two millennia, urged that all inherited assumptions should be questioned. Investigations should begin anew on the basis of principles immune from any possible doubt. This shaking off of orthodox ideas—the "Doctrines of the Schools"—echoed the questioning of church authority that had begun in the Renaissance and Reformation. It began a division within intellectual life: With respect to theology, the traditional methods were favored and the authority of religious scholars was preserved, but for the study of nature, observation and reason were to hold sway.

The pioneers of the new sciences thought that their investigations could disclose systematic patterns underlying the diverse phenomena of the natural world. Those patterns were expressed in quantitative laws: In Galileo's famous phrase, the book of Nature is "written in the language of mathematics." How could the patterns be discovered, and the book read? A popular thought, articulated by Bacon, practiced and adapted by Boyle, and written into the key documents of the early Royal Society, recommended starting with the collection of observations, and generalizing from the connections found in the observed samples. Newton claimed to reject all conjectures, and to base his famous laws on systematic study of the phenomena and careful inferences from them: beginning with observations he saw himself as "making them general by induction" (*induction* being the approved term for inferences that generalize from a sample).

The practice of doing experiments took on a new role. In contrast to the Aristotelian tradition, which regarded ordinary circumstances as revealing natural processes most clearly and interventions as distorting the natural course of things, the new thinkers believed that the special circumstances created by experimentation played a special role in illuminating the underlying structure of the natural world. New quantitative experimental methods encouraged a focus on properties that could be readily controlled and measured. Slowly, a world of qualities and purposes gave way to a world with fundamental properties that were quantifiable and measurable.

New social and institutional organizations were created to support the emerging methods of inquiry. In seventeenth-century England, France, and Italy, societies were founded to foster the exchange of information and to support and evaluate work in the new sciences. These societies began to standardize the form of scientific publications and the professional accreditation of practitioners, creating conditions in which researchers could trust the reports of others beyond their own social circle—a process that eventually led to the institutional structure so crucial to the functioning of professional science today.

We have given a whirlwind sketch, and historians would want to add many details, some of which would recognize continuities between the "new science" and inquiries in the Middle Ages or show how older positions sometimes persisted in the thought of those who most prided themselves on having overcome the past. Yet even if the niceties were acknowledged, two important points for our future discussions would remain. First, what we now think of as science, an enterprise with a distinctive set of approaches and fundamental conceptions, was not always the principal form of inquiry. It replaced an alternative approach to knowledge, focused on different issues—questions about purpose, value, and human salvation. Second, the features of that new enterprise emerged in a particular historical context, in connection with a particular domain of investigation. The pioneers responded to particular deficiencies in the system of knowledge that was taught in the universities of the day, and they elaborated their views of a replacement by beginning with the physics of motion. It is worth pondering those points if you think that science is the best, or even the only possible, form of human knowledge, or if you believe that certain of its features (e.g., reliance on experimentation or use of mathematics) are constitutive of anything worthy of the name "science."

Images of Science

In the chapters that follow, we consider the ways in which philosophers (and sometimes thinkers from other fields: historians of science and sociologists of science, for example) have tried to elaborate general conceptions of science—"images of science" as we call them. Many of those images are popular among scientists (although scientists tend to be unworried about the details of their favorite image), and they are prominent in newspaper discussions of science, as well as in books written for the general public. You are probably familiar with most of them.

The images are often taken to describe how science actually is. When someone (historian, sociologist, or journalist) discovers that a piece of work fails to fit the preferred image, though, there is often a significant shift in perspective. The image is no longer seen as *descriptive*, but as *normative*: This is how science *should be*. Despite this shift, a connection with description usually remains. The problematic work is a deviation from the proper course of scientific activity, a course taken to be exemplified in the overwhelming majority of scientific investigations.

One of the most enduring images, stemming from Bacon and the charter of the early Royal Society, and persisting into the present, views science as a reliable means of accumulating useful knowledge. The scientific method starts with observations, prudently generalizing from them to yield more general conclusions. The patterns that emerge from this activity can be applied to predict and control the course of nature in ways that improve the human lot. Over the years, decades, and centuries, the community of scientists builds a vast edifice of useful knowledge.

A different image, made popular by the influence of the twentieth-century philosopher Karl Popper, opposes the thought of beginning from observation. Scientists inevitably need some guiding idea—they cannot simply obey the command "Observe!" If you were told to observe, you would have to ask just what you were supposed to be observing! They begin from conjectures (the bolder, the better), embarking on a daring adventure—an exploration of territory beyond the frontiers of knowledge. Yet their voyages are tempered by thorough self-criticism. No hypothesis, however plausible or apparently well supported by the available evidence, is beyond criticism. With luck, the adventurers may arrive at useful truths, valuable because they contribute to human goals or because they are worth apprehending for their own sake, yet acceptance of these must always be provisional. Science is to be celebrated both for its delivery of provisional truths and because it is an expression of human striving and human freedom.

In a third image, science is the epitome of rationality, the best way of using our minds to make sense of the world. Although our scientific conclusions are always revisable, there are objective logical relations between evidence (on the one hand) and hypotheses and theories (on the other). So it is legitimate to speak of some of the great scientific achievements—the fundamental principles of quantum mechanics, our knowledge of laws of chemical combination, the molecular basis of inheritance—as objectively well established: not so certain that they are immune to revision, but as close to certainty as we can ever come. Science aims at a systematically unified and complete account of nature, and the history of the past four centuries can be understood as making significant progress toward reaching that goal.

All three images are *celebratory*. They view science as a wonderful thing, perhaps the best activity in which people have ever engaged, the highest expression of humanity. Chapter 2 explores a philosophical tradition, exceedingly influential during the past eighty years, that has taken the celebratory themes extremely seriously and has attempted to articulate them in detail. By considering that tradition, we see how problems and difficulties arise. These problems and difficulties need not shake our confidence in the value of science, but they will suggest that a more subtle and nuanced account of it is in order.

Chapter 3 begins to work out a view of the sciences (plural!) that pays attention to some of their differences, and to consider how the knowledge provided in various areas of research might bear on central philosophical questions about nature and our place in it. In Chapter 4, we begin to consider a fourth image, one that has emerged in the past half-century, sparked by investigations in the history of science.

According to that image, science is a thoroughly human activity, one of many in which people engage. As in other domains of human affairs, the practice of science is shaped by social interaction. Scientific choices are affected by scientists' social context, personal interest, and ambition, and by their broader beliefs about everything, including religion and politics. On some versions of this image science is neither cumulative nor progressive. It is better viewed as a series of exercises in articulating dogma, punctuated by changes in doctrinal fashion.

Chapter 4 identifies the historical considerations that gave rise to this image, and examines its credentials as a challenge to the celebratory images that philosophers have often tried to elaborate. In Chapter 5, we turn to a more radical set of criticisms, taking up the views of people who think that the historical and contemporary practices of the sciences are defective in important ways. Among these are not only many who hope to improve our ways of investigating the natural world, but also some who contest the thought that science is to be celebrated or even cherished.

Our last chapter attempts to synthesize some of the questions and proposals that have emerged, by thinking systematically about what goals the sciences might properly pursue. We argue against the traditional position that science is, or at least should be, a value-free zone. Instead, we suggest, philosophical discussions of science should constantly bear in mind the place that scientific knowledge plays in people's lives, especially in the crafting of highly consequential policies. Questions about the embedding of science in society have often been neglected in the philosophy of science. We take them to be central.

Along the way, we offer brief discussions of the kinds of questions with which we began. Only toward the end, however, when the questions about values are explicitly before us, will we be able to do all of them justice.

Suggestions for Further Reading

For background on controversies about climate change, see Naomi Oreskes and Erik Conway, *Merchants of Doubt* (New York: Bloomsbury, 2010) and Michael Mann, *The Hockey Stick and the Climate Wars* (New York: Columbia University Press, 2012). An excellent introduction to the debates about race and intelligence is given in N. J. Block and Gerald Dworkin (Eds.), *The I.Q. Controversy* (New York: Pantheon, 1976; especially the two-part article by the editors); the findings about subdivisions of the human population are presented in Noah A. Rosenberg et al., "Genetic Structure of Human Populations," *Science*, 298, 2002, 2381–85.

Alexandre Koyré, *From the Closed World to the Infinite Universe* (Baltimore: Johns Hopkins University, 1957) is a classic older study of the Scientific Revolution. Thomas Kuhn's *The Copernican Revolution* (Cambridge, MA: Harvard University Press, 1957) provides an accessible account of the astronomical debates. E. J. Dijksterhuis, *The Mechanization of the World Picture* (New York: Oxford University Press, 1961) is a magisterial presentation of the changes in physics during the sixteenth and seventeenth centuries. Steven Shapin, *The Scientific Revolution* (Chicago: University of Chicago Press, 1996) gives a historically sophisticated discussion of the period that is also extremely clear and readable.

CHAPTER 2

The Analytic Project

Demarcating Science

In Chapter 1, we dodged the question of how to characterize science, settling instead for a list of exemplary sciences. Philosophers have sometimes hoped to do better than this, and there is motivation to try. For no mere list, even a complete one, could identify the characteristics the sciences share. If you believe, as many people do, that science is a thoroughly good thing—the best form of knowledge human beings can achieve and one of humanity's crowning achievements—you might think that recognizing the shared characteristics of the sciences would reveal why science has this special status. So you would be interested in finding a better answer, one that would provide a "criterion of demarcation."

Demarcating science is not just a popular game among philosophers: it pervades media discussions of science, scientists' own reports of what they do, and—especially—the opening chapters of science textbooks. Commentary on science routinely begins with the idea that science has a special method, and this method accounts for the special status of scientific knowledge. Many commentators would link the method to the historical events sketched in Chapter 1, claiming that the seventeenth-century pioneers who launched modern science didn't simply discover particular things about the natural world, fathoming the motions of projectiles and the laws of optics—they also worked out how proper investigations of nature should go. Descartes, for example, believed that his discoveries of coordinate geometry, principles of light transmission, and the explanation of the rainbow were generated by using a common method, one that could be applied far more widely, and he wrote a treatise, the *Discourse on Method*, to explain it to his readers (presumably in the hope that others would use it to make further advances in knowledge).

Although very few people agree that Descartes succeeded in identifying it, faith in the existence of a special *scientific method* is widespread, and it is often supposed that the luminaries of the seventeenth century—Bacon, Galileo, and Newton, as well as Descartes—had some intimations of correct method. But contemporary attempts to say just what scientific method is are typically vague. We're told that scientists "test their hypotheses against observations or experimental results" and "base their conclusions on the evidence". Philosophers of science have hoped to make these suggestive phrases more precise, to say what the proper procedures for testing are and what it is to pay attention to the evidence. Before we can make progress in this enterprise, however, it will be helpful to introduce a pair of distinctions.

The first distinguishes two phases of scientific activity, the *context of discovery* and the *context of justification*. Discovery occurs when someone moves from a state of ignorance on a topic to entertaining the correct answer. Before 1920, nobody knew what physiological condition causes diabetes. Then, one evening, a young Canadian doctor, who happened to read two unrelated articles in close succession, came to a novel idea: Particular structures within the pancreas were crucial to the regulation of glucose, and extracts from these structures could be used to treat diabetic patients. Frederick Banting's work of *discovery* started that evening, but the *justification* of his idea took months of painstaking experiments (see "The Discovery of Insulin").

∽ The Discovery of Insulin

During the late nineteenth century, medical researchers came to recognize that diabetes (type I diabetes, the disease that strikes people in their youth, and that, in those days, usually caused death within a year, even with the best treatment) was the result of malfunctions of the pancreas. That insight inspired attempts to isolate some pancreatic extract that might be given to diabetic patients that would restore their ability to regulate their glucose levels. By the early twentieth century, however, a sequence of failures had convinced the most eminent researchers on the subject that the search for the restorative "juices" was hopeless.

In 1920, Frederick Banting, a young doctor, opened a practice in London, Ontario. With few patients and with educational debts to pay, he supplemented his meager earnings by taking a part-time job at the University of Western Ontario medical school. On Sunday October 20, in preparation for a lecture on carbohydrate metabolism that he was to give to the medical students, Banting read up on the topic. That evening, tired from his studies, he picked up a newly arrived medical journal, and decided that skimming through it might help him to fall asleep. The leading article was on the relationship of a pancreatic structure, the islets of Langerhans, to diabetes. It described a rare occurrence, a pancreatic stone, that had blocked the main duct of the pancreas.

(cont).

Taken on its own, that article only supported a very modest conclusion, but, for Banting, steeped as he was in the current literature on carbohydrate metabolism, it began a new train of thought. Around 2 A.M., he got up and wrote in his notebook:

Diabetus
Ligate pancreatic ducts of dogs. Keep dogs alive till acini degenerates leaving Islets.
Try to insolate the internal secretion of these to relieve glycosurea.
[The spellings are Banting's own.]

Essentially, Banting had discovered—or guessed?—the way to treat diabetes.

Yet, as he was well aware, he needed evidence. Assembling the experimental support took the better part of two years. Banting's first step was to seek the advice of a local expert, Dr. J. J. R. Macleod, who shared the orthodox skepticism about pancreatic extracts. Joining forces with one of Macleod's student assistants, Charles Best, Banting learned how to perform the necessary operations on dogs, and, after a series of difficulties, punctuated by partial successes, the duo of Banting and Best were able to isolate an extract that reliably worked.

In 1923, the Nobel Prize in Physiology and Medicine was awarded to Banting and Macleod (Macleod had presided over the work of Banting and Best, sometimes offering help or advice, sometimes expressing skepticism, and sometimes infuriating Banting). Banting was so angry at the committee's decision that he toyed with refusing the award. He eventually accepted it, publicly announcing that his share was to be divided with Best, and, to the end of his life, he emphasized Best's role as an equal partner in his work.

The most ambitious view of scientific method would suppose that scientists have a systematic ability to generate the answers to the questions that arise for them and to acquire evidence that will justify their conclusions. Descartes believed he had a method of this sort, a key to both discovery and justification, that would unlock the secrets of nature. Historical attention to the episodes in which major new scientific ideas are initially generated suggests, however, that a method of systematic discovery is too much to hope for. Discovery is all too often serendipitous: Banting's chance reading led him to formulate a new idea, just as Darwin's perusal of a book on political economy stimulated him to think of natural selection; Kekulé came to imagine a structure for benzene when he was staring into a fire, and the hot coals rearranged themselves into a striking form. Significant novel hypotheses rarely emerge from a mechanical compilation of data, but often require the imaginative leap that introduces a new conception. If there are methods of discovery, they are not fully general, but limited to domains with particular features (e.g., specific types of statistical data).

Philosophical orthodoxy thus came to abandon any systematic account of discovery, recognizing the roles of chance and imagination in the creative work of generating hypotheses. Scientific method belongs in the context of justification. Once the investigator has thought up the hypothesis, by whatever means, method comes into play in the activity of assembling evidence and accepting only those ideas that are well supported by the evidence. The sciences are distinguished from other attempts to achieve knowledge because their conclusions are firmly established.

To develop this proposal, the second distinction is needed, a distinction between the ways justifications work in different areas of inquiry. Descartes published his *Discourse on Method* as the preface to three other *Discourses,* one devoted to mathematics (coordinate geometry) and the others on optics and meteorology. Yet, on the face of it, the methods of justification in mathematics and in the natural sciences are very different. Very few thinkers have maintained that our justifications for believing simple arithmetical identities are provided by our everyday encounters with small groups of objects: Most have supposed it ridiculous to hold that our knowledge that $2 + 3 = 5$ is based on experiences of combining beans or pebbles. Some philosophers propose that there are other sources of knowledge, independent of observation: Mathematics, including arithmetic, geometry, "higher" mathematics, and logic, consists of "truths of reason," although it is notoriously difficult to explain precisely just what reason is, and how it operates to yield mathematical knowledge.

One popular view that tries to understand logical and mathematical knowledge, without engaging in speculative psychology by introducing special faculties of reason or intuition, contends that there are fundamental principles that underlie all thought, the basic axioms of logic and elementary rules of inference. All the truths of logic can be obtained from the basic axioms using the elementary rules. With the aid of definitions, a wider class of truths—the *analytic truths*—can be generated. Analytic truths ("Bachelors are unmarried males" to take a simple, much-used example) are simply true statements that can be seen as logical truths when defined terms are replaced with the language of their definitions (*bachelor* is defined as *unmarried male,* so that the original statement becomes "Unmarried males are unmarried males"). Using this notion, it is possible to explain the status of mathematics: Mathematical truths are analytic—mathematics is disguised logic. Or, adopting a different but related approach, it might be supposed that mathematics is the study of formal systems, languages whose nonlogical expressions are strictly meaningless, and mathematical work consists in developing these languages and identifying the theorems that can be proved in them.

If either of these proposals is accepted, mathematical justification is separated from justification in the natural sciences. Mathematics can proceed sure-footedly without ever looking at the physical world, proving its conclusions in ways that secure them against future revision. By contrast, the natural sciences are not just collections of analytic truths (nor are they merely formal systems). Hypotheses in the sciences must be based on observation of nature, and, as will quickly become apparent, this means that scientific justification is not a matter of strict proof.

People do sometimes talk of scientific proof—often, when they want to dispute a claim, by suggesting that "proper scientific proof" is lacking. Behind this language is a reasonable thought. Mathematics has first principles—axioms, or perhaps ultimately the fundamental laws of logic—and the mathematical proof of a theorem consists in deducing it from the axioms. In parallel fashion, science is based on statements that can be immediately and directly justified by observing parts of the natural world—perhaps parts that have been specially prepared in laboratories, perhaps parts that are untouched by human manipulations. The scientist justifies a hypothesis by using these basic statements as the first premises in inferences that accord with the rules of an extended logic, one that covers the crucial nondeductive arguments that are used to support scientific hypotheses and thus includes the principles of valid deduction and more besides. Scientific method consists in assembling basic statements and constructing arguments sanctioned by this extended logic to ground conclusions about nature.

Two obvious questions now arise. First, what are the proper starting points—the basic statements grounded in immediate observation—for justificatory arguments in science? Second, what types of inference, beyond deduction, should be allowed in an "extended logic" for science? Both questions prove very difficult to answer.

In reporting their observations of the world, people typically draw on their background knowledge. We talk about seeing happy children, or lost dogs, or dangerous intersections. Scientists are no different. The technician who sits in front of a cloud chamber reports the emission of a positron, the bacteriologist observes the presence of an antibiotic-resistant strain in the cell culture, and the primatologist notes down a dominance display by the alpha male. Are these the sorts of statements that should serve as the bases of scientific knowledge? It seems not, for they already take for granted large chunks of accepted scientific theory, and these could be mistaken. So perhaps the real grounds of science lie in more minimal claims, reporting that there was a track of a particular shape across a particular screen, there were clumps of opaque patches in a dish, or an animal engaged in specific sorts of movements. If claims like these are to support the sorts of hypotheses in which particle physicists, bacteriologists, and primatologists are typically interested, they will have to be supplemented with vast numbers of further reports about the construction of the apparatus used, the circumstances of the observation, and so forth—all formulated in the same minimal way. How exactly this is to be done is far from clear. Moreover, even though the retreat to more cautious observational reports guards against importing errors from parts of science that are taken for granted, there are other risks of going astray. Can the technician be sure that the streak on the screen had a specific, precise shape? Or that the apparatus was properly constructed and maintained? If the goal is to eliminate all kinds of doubt, it seems that the bases for scientific knowledge must be even more minimal—perhaps just reports of bits of sensation; patches of color and bursts of sound—but this means they will be even further removed from the superstructure they are supposed to support.

The difficulties with the extended logic allegedly used in justifying scientific hypotheses and theories are even easier to appreciate. Many of the most impressive claims in the physical sciences are general: Newton's law of universal gravitation asserts that there will be a force of attraction between *any* two bodies; chemical texts contend that combustion *always* involves the absorption of oxygen. Yet the observational evidence anyone—or everyone—has, or could have, is finite. The vast majority of pairs of bodies, or of episodes of combustion, will never be observed. As Hume recognized over 200 years ago, it is logically possible—that is, consistent with the principles of deductive logic—that one or more of these unobserved instances could be at odds with the generalization. Indeed, the history of science reveals numerous occasions on which the accumulation of evidence forced scientists to revise generalizations that had previously been confidently accepted: Mendel's simple rules for the assortment of traits have been revised to accommodate all sorts of genetic mechanisms that nobody had suspected; Newton's extraordinarily successful laws of motion had to be refined by quantum mechanics and by the special theory of relativity. So the extended logic has to abandon the notion of validity central to deductive logic. If an inference is deductively valid, it is impossible for the premises to be true and the conclusion to be false. Scientific inferences cannot be like that: for, when you infer from a sample to a broader population, it is always possible for the premises (the descriptions of the sample) all to be true and the generalization about the population to be false.

Science is *revisable*. Hence to talk of scientific "proof" is dangerous, because the term fosters the idea of conclusions that are graven in stone. Does recognizing the openness of present science to future correction undermine the idea that science is the best form of knowledge about the natural world? Not decisively. Perhaps, even if scientific conclusions are not so firmly supported that they are immune from revision, they are still justified more thoroughly than other things we take ourselves to know. It should be apparent, however, that identifying the sources of superiority is going to be a complicated business. Moreover, it may be quite legitimate to wonder if other areas of human knowledge do just as well as the sciences, or to take a more positive attitude toward some of our commonsense lore or the claims made in parts of the social sciences and the humanities.

Perhaps the very revisability of scientific hypotheses and theories is what sets them apart. This idea has inspired the most influential way of attempting to demarcate science and to explain its superiority. Aside from logic and mathematics, all our beliefs may require modification in the light of future evidence. The distinctive feature of the natural sciences is that they take this predicament seriously and face up to it with intellectual honesty. Instead of viewing scientific claims as especially well supported, we should conceive them as made in an ongoing enterprise that is rigorously concerned with the eradication of error.

This position, eloquently presented and defended by Karl Popper, is prominent in the public statements of scientists, particularly at moments when parts of science are on trial. The mark of real science, it is claimed, is *falsifiability*. A hypothesis is said to be falsifiable just in case there is some piece of evidence that could conclusively refute it. In counting scientific hypotheses as falsifiable, you do not suppose that they are false—indeed, scientists usually hope that their falsifiable hypotheses are true—but you pay them a compliment. Good hypotheses (scientific hypotheses) stick their necks out; they take risks by making claims that nature could reject. Good scientists expose their pet ideas to the judgment of nature, attempting to subject them to the most rigorous tests. When the test results are negative, they abandon even their most cherished claims. This openness to correction (it is often said) is the hallmark of science, and the source of the superiority of science over other kinds of human knowledge.

One of Popper's original examples can help to motivate this falsifiability criterion of demarcation. Skeptics about Freudian psychoanalysis have often viewed it as merely pretending to be scientific, a game without rules. The source of the worry is the apparent ability of psychoanalysis to cope with any evidence whatsoever; to find supporting instances everywhere. For example, at one stage of his analysis of the "Rat Man" (see "Freud on the Rat Man"), Freud ventured the prediction that his patient had been punished for some kind of infantile sexual offense (probably masturbation). A later discovery of an episode in early childhood, where the little boy had reacted quite violently to punishment, revealed that the bad behavior had nothing to do with sex. Freud then modified his position to suggest that, in the patient's consciousness, the incident had later been "sexualized"—the connection with sexuality remained, but was located in the patient's mind rather than in the historical event. From the Popperian perspective, the availability of such methods of accommodation exposes the fact that the original hypothesis about the causes of the Rat Man's behavior was unfalsifiable—and thus unscientific.

∽ Freud on the Rat Man

The "Rat Man" (Paul Lorenz) sought treatment from Freud for an "obsessional neurosis," in which he suffered from recurrent thoughts that people he loved or admired would suffer excruciating agony. The condition came to a head during a period of military service, when Lorenz had encountered a sadistic officer, who had described a particularly horrible form of punishment in which a pot full of rats was applied to the victim's buttocks, so that the rats bored their way into the anus. After hearing of this torture, Lorenz could not resist thinking that it was to be inflicted on the "lady" whom he admired and on his father (who was already dead).

Freud's case history reconstructs a fascinating web of associations that he uses to interpret Lorenz's behavior. Freud traces the obsessive thoughts to a conflict with the patient's father, begun in childhood, and eventually hypothesizes that "when he was a child of under six he had been guilty of some sexual misdemeanor connected with onanism and had been soundly castigated for it by his father." Lorenz then reported that his mother had often told him of an unusual episode in his childhood, when his father had beaten him for doing "something naughty," and he had become enraged, calling his father all the common names he could think of. (Because his vocabulary was innocent, he had screamed, "You lamp! You towel! You plate!".)

Lorenz had been three years old at the time, and, according to his mother, he had been punished because he had bitten someone. Freud notes this fact in his case history, and adds a long footnote. In that note, he suggests that "childhood memories are only consolidated at a later period, usually the age of puberty." Remembrances of personal history are reconstructed in the ways nations arrive at legends about their past, so that worrying forms of sexuality can be washed away or transformed. Alternatively:

> It becomes clear that in constructing phantasies about his childhood the individual *sexualizes his memories*; that is, he brings commonplace activities into relation with his sexual activity, and extends his sexual interest to them.

So Freud provides two strategies for coping with the unwelcome report that the punishment and the ensuing temper tantrum were precipitated by the little boy's biting someone. Perhaps the naughtiness was indeed sexual—the

(cont).

father caught young Paul masturbating—and the mother sanitized the affair. Or perhaps the report is fully accurate, and, as his memories of childhood "consolidated," the adolescent Lorenz associated the punishment with interference in his sexuality, and so came to conceive the conflict with his father as sexual. Freud's willingness to use these strategies (especially to use both in the same case, covering every eventuality) seems to put him at odds with Popper's image of the rigorous—"intellectually honest"—scientist, who willingly abandons his hypothesis in the face of evidence that controverts it.

Science is supposed to be special because the fundamental rule of practicing science insists on intellectual honesty. Evidential support that is gained simply by constructing hypotheses that are vague or adjustable breaks this rule. Scientists must specify in advance observational or experimental findings that would show their proposed hypotheses to be false—and then follow through by giving up those hypotheses when the test results prove negative. But an insight of the French historian and philosopher of science, Pierre Duhem, casts doubt on the extent to which the concept of falsifiability can be applied to individual hypotheses. As Duhem noted, hypotheses are tested "in bundles." Whenever researchers perform experiments, they have to rely on assumptions about the apparatus they use, and the background conditions that are necessary for it to function. (If this were not so, then all sorts of major scientific generalizations would routinely be falsified by students in high school and university laboratories; students are usually not hailed for their groundbreaking discoveries, but told to try the experiment again until they "get it right.")

Copernicus's proposal that the Earth has an annual motion around the sun is often taken as one of the great hypotheses in the history of astronomy. In the late sixteenth century it was not popular, at least as a literal account of planetary motion. Indeed, it would seem to have been "falsified"! For if the Earth moves around the sun, then distant stars (the "fixed stars") should appear at different angles at different seasons of the year, just as spectators are viewed at different angles by children riding on a carousel. But they do not—hence the name *fixed stars*. Galileo, one of the few early adherents of the hypothesis, responded to the charge of falsification by pointing out that the implication of vision at different angles would not hold if it were supposed that the stars were far enough from the Earth; for then the angular shift would be too tiny to be observed. Virtually none of his contemporaries had thought of a universe with such vast distances, and Galileo could not provide any independent measurement in support of his idea. Yet he proposed to "save" the Copernican hypothesis by amending a traditional assumption about the size of the cosmos.

Was that legitimate? Indeed, was it any different from Freud's response to the details about the Rat Man's early life? If not, then one of the greatest advances in early modern science was made by a thinker who defied Popper's canons of good scientific procedure. If, however, we are more sympathetic to Galileo (and maybe also to Freud), then the simple Popperian story of falsification and falsifiability must be abandoned. Tests of a central hypothesis are conducted by introducing auxiliary assumptions, and it is—at least sometimes—quite reasonable to question the auxiliary assumptions rather than jettisoning the central claim. How far can this go? Could a scientist always divert the blame to some auxiliary hypothesis, and thus hang on to a pet idea? That seems like exactly the nonscientific attitude Popper hoped to expose, but it is not easy to see how to draw the relevant distinctions.

As we already noted, appeals to falsifiability are popular when science is on trial. In the past decades, creation scientists have been told by the courts that their proposals do not belong in the high school curriculum, even as potential alternatives, because those proposals are not falsifiable and therefore not scientific. If the notion of falsifiability is intended to apply to individual hypotheses, then a very significant range of major scientific proposals—the Copernican hypothesis, Newton's law of universal gravitation, the thesis that matter is composed of atoms, and many more—will be ruled out. So the only sensible version of the demand for falsifiability is to apply the criterion to whole *systems* of hypotheses ("bundles," as Duhem called them). Yet it is now trivial for creationists to conform to the demand. They can draw up a list of potential observational findings that they have good grounds for believing either to obtain or to be very hard to refute in practice. Their "system" can then be adjusted to incorporate hypotheses to the effect that if the history of life began some 10,000 years ago with an act of divine creation, just those observable results would be found. Hence there are potential observations—findings contrary to those anticipated—that would falsify the system. Further, if luck ran against them, and contrary observational reports were to be validated, then creationists could again adjust, discarding some "misguided auxiliary hypotheses" to protect the core of their system. They could even point to Galileo's good example.

Our search for a criterion of demarcation has failed. Yet the motivation with which we began, the urge to understand the distinctive features of science that make scientific knowledge seem especially reliable and important, remains. Even if we suspect that the natural sciences are continuous with other parts of human knowledge, that common sense and social science and the humanities sometimes do just as well, it is worth considering what features distinguish our inquiries at their very best, and turning to the sciences for guidance in making those features clear and explicit. By doing so, it might be possible to see how to make struggling fields go better: If we knew more about what has made for good investigations in chemistry (say), there might be lessons for areas where reliable knowledge would be very welcome—the economics of global markets, for example. Surely, there are many domains of investigation that lack the striking qualities of the natural sciences at their best, domains in which nothing is ever settled, in which loose and cloudy talk predominates, and that seem to lurch from one fad to another. Appreciating the scientific virtues might be the prelude to valuable reform.

The philosophy of science of the past half-century (and more) has been dominated by ventures toward this goal. The *analytic project*, as we call it, has aimed to provide general accounts of three main characteristics of the natural sciences (see "Philosophical Sources for the Analytic Project"). The thought that the sciences obtain especially sound knowledge has inspired the search for a *theory of confirmation* that would delineate the general ways in which scientific hypotheses are supported by evidence. The thesis that scientists strive for clear and precise formulations of their ideas has sparked attempts to provide a *theory of theories and theoretical language.* Finally, the view that the sciences deepen our understanding of the natural world has given rise to proposals for a *theory of explanation*. These ventures within the analytic project are considered in the next three sections.

✆ Philosophical Sources for the Analytic Project

Philosophical attention to science in the early twentieth century took inspiration from the new mathematical logic. At the end of the nineteenth century, several mathematicians, most notably Gottlob Frege, introduced formal languages with the intention of clarifying the logical structure of mathematical proofs, and the work was further advanced by Bertrand Russell and Alfred North Whitehead in their influential three-volume work *Principia Mathematica*. The new logical tools stimulated the members of the Vienna Circle (a group of young philosophers trained in mathematics and science) to envisage a long overdue "correction of philosophy." They lamented the inability of traditional philosophers to settle their differences and to achieve cumulative progress in the ways so conspicuous in the history of mathematics and of the natural sciences. They diagnosed the failure of philosophy as lying in the fact that philosophical debates either focused entirely on questions for which no resolution was in principle possible, or else mixed genuine controversies with issues that cannot be settled. Traditional philosophy was a mixture of pseudo-problems and real questions. The proper task of philosophy was to expose the pseudo-problems, and to liberate the serious issues in a way that would prepare for their solution.

(cont).

Philosophy goes astray when its language fails to be connected in any appropriate ways to the basic observational vocabulary. When language cuts loose from its moorings, there may be the appearance of sensible discourse, but the sentences uttered turn out to be cognitively meaningless. The reformers proposed to elaborate a *criterion of demarcation* that would distinguish meaningful discourse (science) from empty blather, by saying precisely what is required for statements to be connected in the appropriate way to observational terms. This criterion was to be formulated in terms of the new logic.

The reformers' early program, styled as *logical positivism*, was ambitious and optimistic. Logical positivists began with the thought that all meaningful language should be explicitly definable in terms of observational vocabulary, but they soon abandoned this idea—for they did not see how to provide explicit definitions for key terms of theoretical physics. Instead, they suggested a simple criterion: Statements are cognitively significant just in case they are conclusively verifiable. But anything so simple would debar scientific generalizations (as Hume had shown, long before). So the initial version gave way to the proposal that statements have cognitive content if they entail, by the principles of deductive logic, statements in the observational vocabulary. This is an initially plausible idea. For the scientific claim that bodies released near the Earth's surface accelerate toward it entails that it will not be the case that a body is released and flies upward (to cite just one consequence). On the other hand, the thesis that the Idea of Spirit is Freedom (just the kind of grand metaphysical claim positivists saw as endemic to misguided philosophy) seems to lack observational implications.

Positivism's conscientious commitment to elaborate its ideas as precisely and clearly as possible, drawing on the resources of formal logic, revealed that, for all its plausibility, this proposal will not do. Let G (for Galileo) be the scientific hypothesis mentioned in the last paragraph, and H (for Hegel) be the metaphysical claim. Consider now the statement $G\&H$. Because this statement deductively entails G, it entails all consequences of G as well. Among those consequences are some observation statements, for G is a respectable scientific hypothesis. Hence, by the criterion, $G\&H$ is cognitively significant. Presumably, then, traditional philosophers can rewrite their works by simply adding a few scientific statements at critical places.

Worse is to come, for trouble threatens from the opposite direction. Consider Newton's law of universal gravitation: Every body attracts every other body with a force that is directly proportional to their masses and inversely proportional to the square of the distance between them. You might initially think that this has observational implications about the potential motion of bodies. Unfortunately, any claims about the motions depend on what *other* forces are acting. It is, of course, true that, if the Newtonian law is

(cont).

supplemented with other statements—*auxiliary hypotheses* that specify that no other forces are acting, or the character of such other forces—then conclusions about motions are derivable. To allow that the law counts as cognitively meaningful on these grounds would, however, reduce the criterion of cognitive significance to toothlessness. For the proposal would now be as follows: *N* is significant because *N&A* has observational consequences. Would-be metaphysicians could rejoice in the reformulation: They, too, can find auxiliary hypotheses to generate observational consequences. Suppose *O* is the observational statement "The rat pressed the bar once." Let *A* be "If *H* then *O*." *H&A* has *O* as a consequence, so, by the revised criterion, *H* counts as cognitively significant.

Such difficulties prompted more refined versions of the criterion of cognitive significance, and the more refined versions encountered new problematic cases, until by 1950, the repeated troubles provoked a change of agenda. Positivists concluded that any sharp demarcation was hopeless. Instead, they saw their task as specifying, as clearly and exactly as they could, the principal features that underlay success in those areas of science that seemed to make the most spectacular progress. If the distinctive attributes of confidently advancing fields of knowledge were understood, then, perhaps, other areas of inquiry might benefit from the examples. Parts of the social sciences that often seemed bogged down in endless disputes about fundamentals might acquire more reliable methods. Reform might spread further, perhaps even into philosophy.

Three attributes of science seemed particularly salient: scientific hypotheses are well-confirmed, scientific theories are precise and clearly formulated, and science gives genuine explanations. So there seemed to be three crucial tasks, those of analyzing the concepts of confirmation, theory, and explanation. The analytic project was born, and there was a change of name. *Logical positivism* was replaced by *logical empiricism*.

Confirmation

As we have seen, scientific evidence falls short of conclusive proof. Not only the largest generalizations of theoretical science, the equations that govern gravitation or electromagnetic fields, but also relatively humble extrapolations from the properties found in a sample, are impossible to establish definitively. General hypotheses are subject to revision. Nevertheless, it is natural to suppose that sampling has a point. If we base a conclusion about all members of a population on an investigation of a sample drawn from that population, we are better off than if we simply guess. We think of ourselves as acquiring greater justification by investigating some of the cases to which the generalization applies. Conclusions about the effects of drugs are tested by seeing what happens when they are fed to laboratory animals, and eventually to people, and, once enough of this has been done, government agencies announce decisions about whether these drugs are safe and effective enough to use. The verdict remains fallible, as tragic examples of unanticipated side effects should remind us, but the lengthy process that precedes it provides evidence in support of allowing the drug to be distributed. It does not prove, but it *confirms*. How exactly?

The Hypothetico-Deductive Method

One popular answer involves what scientists call *strong inference* and philosophers dub the *hypothetico-deductive method*. From a hypothesis (or a cluster of hypotheses) you can derive many statements about what would be detected in nature if the hypothesis (or the cluster) were true. Scientists choose some of these statements—*test implications*, as they are often called—and investigate the world to find out if they actually hold. Suppose that a number of test implications are explored, and each of them turns out to be correct. Despite the fact that these aggregate results do not prove the hypothesis, they could be held to confirm it. From a logical point of view, the structure of the situation is as follows:

If H then $O_1, O_2, \ldots, O_n$
$O_1 \& O_2 \& \ldots \& O_n$
So: H.

The inference here is not deductive—indeed, it is a well-known deductive fallacy—but many people claim that it is a cogent nondeductive inference. By this they mean that, although it does not measure up to the standard of deductive validity, which demands that it cannot be the case that the premises are true and the conclusion false, this inference nonetheless provides some support for the conclusion. One very natural way to express the thought is to declare that the investigation of the various test implications has put the hypothesis at risk: It has been subjected to a process in which its flaws might have been exposed. That it has survived should count in its favor.

Even if the idea is natural, it remains vague. Nothing has so far been said to explain what kinds of test implications confirm a hypothesis, and when they add up to evidence that is strong enough to warrant (tentative) acceptance. Thinking about the example of drug testing enables us to appreciate differences in the force of the evidence provided by alternative ways of exemplifying the basic approach. Responsible agencies explore the effects of drugs on a wide variety of animals. They might think that mice are good representatives for all mammals (including human beings), but they are not likely to think that a sample of cases, entirely composed of female mice of a particular age, fed exactly the same diet, and given the drug at precisely the same developmental stage, would provide particularly good evidence for safety. If the hypothetico-deductive method is to be deployed to yield serious confirmation, a more diverse collection of test subjects would be needed. In practice, experimentalists decide how to construct good trials by using quite specialized knowledge about the animals they plan to use. So they might conclude that a sample composed entirely of mice would be representative, but that the membership needs to be varied by sex, age, and other biological factors. How are decisions like this made?

Here is one way of understanding the situation. Scientific research always operates against the background of a field of possibilities. In light of what previous investigations have shown, it is reasonable to think that certain kinds of variation in mice might be relevant to the effects of a drug, whereas others might not. There may be reason to think that males and females will react differently, but that obese mice and thin mice are not likely to do so. If that is so, it will be important to have both males and females in the study, but the relative svelteness of the subjects can be ignored. So, we might suppose, the background state of physiological knowledge provides a list of potential confounding variables, and the testing of the hypothesis requires considering all possible combinations of them. In terms of the schema offered for strong inference, each combination of the listed variables must figure in at least one of the reports, $O_1, O_2, \ldots, O_n$.

Viewed in this way, the scientific investigator is in something like the position of the detective investigating a murder in a remote country house (in the classic scenario from mystery fiction). Background knowledge suggests that the villain cannot have entered from the outside, and so must be one of the people known to be present. If the detective can run through the list of suspects, showing that all but one could *not* have committed the crime, then the identity of the culprit can be established, without any need for unjustifiable inferences. As Sherlock Holmes remarks, "When you have eliminated the impossible, whatever remains, however improbable, must be the truth." By the same token, if an experimentalist can reasonably hold that a particular list of variables includes all those that would be pertinent to the failure of a generalization, if all the possibilities are systematically explored and the generalization survives, then the generalization is well confirmed by the inquiry.

If the hypothetico-deductive method is conceived in this way, much of the work of providing justification is done by the background claim about the field of possibilities. The detective's inference is only as good as his assumption that no outsiders could have been involved, and so too with the experimentalist. Often, scientists may feel confident that they are in a good position to specify the genuine alternatives. In the *Origin of Species,* Darwin proceeds by pitting his hypothesis that all organisms are linked in a single tree of life, a nexus of descent with modification, against the hypothesis that individual species have been created independently. He reduces the possible hypotheses to these two by appealing to the background knowledge of the time. His predecessors had used the fossil record to show that, at different stages of the Earth's history, different groups of organisms had lived on our planet. As they and Darwin saw, this left only two alternatives: Either the later types were modified descendants of the earlier ones, or there had been successive waves of species creation.

In many instances, however, especially in newer fields of research, scientists are not in this fortunate predicament. Perhaps they have some ideas about which factors might be relevant to the hypotheses they are considering, but they understand the limits of their background assumptions, and foreswear any claim that the list of factors they can identify would be complete. Under such circumstances, how does the hypothetico-deductive method proceed? Furthermore, even when confidence about the field of possibilities is entirely reasonable, it is worth asking how that has come to be. Presumably, there were prior investigations that led earlier scientists to understand that sex but not relative obesity could be relevant to this kind of physiological investigation. Just how was that judgment justified? Maybe through earlier uses of strong inference, but, in that case, if those inquiries were carried out with a warranted set of possibilities, we should repeat the question. At some stage, we shall apparently come to empirical procedures that do *not* make use of reliable background knowledge—as the thought about the foundations of the Earth goes, it cannot be "turtles all the way down." There must, then, be other modes of nondeductive inference through which scientists can work their way to the happy state where they can rely on background knowledge to limit the field of possibilities.

These other modes of inference are most prominent in fields that struggle to reach agreed-on conclusions. When areas of inquiry are in their infancy, they often seem to make use of inferences that are unguided by any sense of the variables to be considered in sampling. *Blind induction,* as we might think of it, consists in exploring a population defined in some way that is not yet known to lend itself to fruitful generalization, without serious clues about what kinds of variation in the population might be relevant. The prehistory of our most successful sciences was most likely a series of ventures in blind induction, observation of large numbers of members of a particular group in the hope that a generalization would ultimately emerge. As the numbers increased and the generalization remained stable, so too did confidence, until, eventually, the hypothesis became available as a piece of background knowledge. Surely, however, the failures were far more numerous than the successes. The attempt to give a somewhat more exact conception of the hypothetico-deductive method thus exposes the need for a more basic form of inductive inference that enables the method to get off the ground.

As the fields of science advance, their languages develop, furnishing concepts that researchers think are apt for formulating generalizations. Sometimes this involves redrawing boundaries that earlier generations had accepted: Chemistry made significant progress when it refined earlier usages of terms like *element* and *salt*; biology went forward by deciding that whales and dolphins should be grouped with other mammals and not with the fish. Often, these linguistic and conceptual shifts respond to the failure of generalizations—if you think that earth, fire, air, and water are the elements you will have trouble in making successful claims about "all elements." In accordance with the suggestion that hypothetico-deductive inference reaches its mature form by way of an earlier stage of blind induction, in which the necessary background knowledge is accumulated, we might suppose that our ancestors began the practice of generalizing by simply extrapolating from the properties found in samples corresponding to whatever categories they were initially inclined to adopt. As this practice proceeded, people found that generalizations involving some of their original concepts broke down, whereas others were sustained, and they refined their vocabularies in light of these discoveries. Their language and their practices of generalization evolved together, eventually reaching their present forms. We are no longer tempted to generalize using certain kinds of ("monstrous") predicates, because the successful generalizations of the past provide us with what we take as background knowledge, and the background knowledge excludes the "monsters" from the field of possibilities.

Our outline account of how hypothetico-deductive inference works is neither precise nor formal: It bears a closer resemblance to older traditions in thinking about evidence and argument than to the mathematical deductive logic that was introduced at the end of the nineteenth century and that has come to define what logic is. If this is all that can be done for nondeductive inference, then it appears that any dreams of an "extended logic" of confirmation are doomed to disappointment. We do not see how to make some important constituent concepts precise or formal: The selection of adequately diverse samples, the responsible delineation of the background set of possibilities, and the recognition of particular languages as apt for stating generalizations all look like matters resistant to formal definition. Furthermore, there is an additional complication that has not yet been considered.

Although it is easy to think that a finding that accords with a hypothesis supports that hypothesis, there are occasions on which instances actually undermine a generalization. Consider the hypothesis that all human beings die before they reach the age of 200. Well-authenticated cases of people who survive ten, or even twenty, years beyond their first century are untroubling. Imagine, however, that we discover, in some remote part of central Asia, an old man who has lived his entire life on local yogurt, and, after probing his history, we ascertain that he is indeed 198 years old. During the course of our research, however, he succumbs, shortly after celebrating his 199th birthday. This venerable fellow (call him Boris) is, strictly speaking, an instance of the generalization that all humans die before reaching 200. Does Boris add extra support to the generalization? Far from it! After our encounter with him, we would reasonably doubt whether the generalization is correct.

Why is that? Partly because the generalization obtained support not only from records of human lifespans but also from physiological understanding of the ways in which our organs are likely to fail as we age. Boris, we might propose, adds a tiny extra piece of direct confirmation in the blunt way of accumulating instances, simultaneously casting severe doubt on the theoretical considerations that serve as the more important grounds on which our beliefs about the limits of human life rest. That is not the whole story, however. After meeting Boris we come to reappraise the way in which the generalization was previously framed. Knowing that bodies decay and that people die, before or shortly after they reach 100, we wanted to set a bound that would make the generalization safe. We knew that 120 would be too low, and that 130 might be risky, even though no documented case of someone's living to 130 was at hand, but adding seventy extra years seemed to provide room enough to cover even the most fortunate combination of relevant factors. Yogurt-eating Boris shows that our margin was too small.

Confirmation and Probabilities

A very natural response to the fact that scientific evidence falls short of conclusive proof is to invoke the concept of probability: When a hypothesis passes a test, it should count as more probable than it was before. The most promising contemporary attempt to develop a precise account of confirmation (*Bayesianism*) elaborates this idea.

Medical practice supplies examples of how to think about confirmation in terms of probabilities. Suppose you go to the doctor with some bothersome symptoms. Your physician explains that it is probably something quite minor, but your complaints are consistent with an exotic and rare disease. She proposes that some blood be drawn, so that the standard test for the disease can be run. A few days later, you learn that your test was positive. Should you be worried?

Not necessarily. Suppose that the disease is very rare; only one person in 100,000 has it. Assume also that there is a significant rate of false positives: In every hundred people who do not have the disease but who take the test, two will show the signs of having the disease. There are no false negatives: If you do have the disease the test will invariably show it. With this information, it is possible to calculate the probability that you have the disease given that your test was positive: It turns out to be approximately 0.005 (see "Bayes' Theorem and Medical Testing"). So your chance of having the disease has gone up considerably—from one in 100,000 to about one in 2,000—but it remains very low. Notice the crucial role played by the rate of false positives: If the test is quite likely to give a positive result for people who do not have the disease, it can only provide weak evidence for the presence of the disease.

ᗌ Bayes' Theorem and Medical Testing

First, we introduce some notation and terminology. $Pr(A)$ is the probability of A, where A is an event, or the probability that A is true if A is a statement. (So, if A is "The card drawn at random from a standard deck is a spade," $Pr(A)$ is 0.25.) $Pr(A|B)$ is the probability of A given B, or the probability that A is true if B is true. (So if A is "The card drawn at random from a standard deck is a spade," and B is "The card drawn is black," $Pr(A|B)$ is 0.5.) $Pr(A)$ is often called the *prior probability of A*. $Pr(A|B)$ is the *conditional probability of A on B*.

Second, let's explain Bayes' theorem. The precise definition of conditional probability, $Pr(A|B)$ is:

$$Pr(A|B) = Pr(A\&B)/Pr(B)$$

Similarly, $$Pr(B|A) = Pr(B\&A)/Pr(A)$$

and hence, $$Pr(B\&A) = Pr(A)Pr(B|A).$$

Because $A\&B$ is logically equivalent to $B\&A$, $\quad Pr(A\&B) = Pr(B\&A)$.

It follows that $\quad Pr(A|B) = Pr(A)Pr(B|A)/Pr(B)$.

Notice that B is logically equivalent to $[(B\&A)$ or $(B\&-A)]$, where $-A$ ("not-A") is the denial or negation of A. The theory of probability therefore allows us to rewrite $Pr(B)$: $Pr(B) = Pr(B\&A) + Pr(B\&-A)$.

We can now obtain an important form of *Bayes' theorem* (named for the eighteenth-century clergyman, Thomas Bayes):

$$Pr(A|B) = \frac{Pr(A)Pr(B|A)}{Pr(A)Pr(B|A) + Pr(-A)Pr(B|-A)}$$

This theorem is just what is needed to handle cases of medical testing.

Suppose A is "You have the disease," and B is "You test positive." Precisely what you want to know is $Pr(A|B)$, the probability of having the disease if you test positive. The value of $Pr(A)$—the prior probability of having the disease—is the base rate in the population, given as 0.00001. Correspondingly, $Pr(-A)$ is 0.99999. $Pr(B|A)$ is the probability of testing positive if you have the disease, and that was specified as 1 (there are no false negatives).

(cont).

$Pr(B|{-}A)$ is the rate of false positives, known to be 0.02. When you plug these values into Bayes' theorem and do the arithmetic, the value of $Pr(A|B)$ is 0.000499.

    Why is it a mistake to think your chances of having the disease are 0.98? In the standard terminology (jargon), that value comes from neglecting the base rate: People who reach such gloomy conclusions don't take into account the fact that the disease is extremely rare. Although only two people in one hundred who are tested will wrongly test positive, the overwhelming majority of those who take the test will not have the disease (because it is so rare), and that means that a very large majority of those who test positive will be disease-free. Bayes' theorem tells us how to translate this qualitative fact into quantitative terms.

How does this apply to confirmation in the sciences? Let us assume that scientific hypotheses can be assigned probabilities, and that these probabilities measure the extent to which someone should believe the hypotheses. Probabilities are degrees of belief, ranging from 0 (absolute conviction that the hypothesis is false) to 1 (absolute conviction that it is true). To explain this idea of precisely measurable degrees of belief, Bayesians suppose that degrees of belief correspond to your willingness to bet. So if your degree of belief in H were ½ you would be happy with an even bet either for or against the hypothesis; if it were ¾ then you would be prepared to stake $3 on H for the chance to win $1 if you were right. As you learn things about the world, you update your degrees of belief (in accordance with Bayes' theorem).

A famous example from the history of science can show how this is supposed to work. In 1818, the French Academy of Sciences held an essay competition on the topic of the nature of light. Augustin Fresnel submitted a memoir presenting and defending a wave theory of light. During the evaluation, one of the judges, the mathematician Simeon Poisson, recognized that, on Fresnel's theory, there would be a bright spot at the center of the shadow cast by a small circular disk, a result he thought absurd. Another judge, François Arago, was prompted to investigate, and discovered that the shadow did indeed contain a bright spot. That discovery was taken to provide dramatic confirmation for Fresnel's hypothesis. Bayesianism can tell a plausible story about what occurred in this episode: If the bright spot was absurdly improbable unless Fresnel's theory was correct, its appearance required scientists to increase their degrees of belief in the theory dramatically (see "Bayes' Theorem and the Wave Theory of Light").

◡ Bayes' Theorem and the Wave Theory of Light

Let F be Fresnel's version of the wave theory of light, and S be the statement that there is a bright spot at the center of the shadow of a small disc. We want to know $Pr(F|S)$.

Once again, Bayes' theorem enables us to compute this value, provided we know certain other probabilities.

$$Pr\,(F|S) = \frac{Pr\,(F)Pr\,(S|F)}{Pr(F)\,Pr(S|F)+Pr(-F)Pr(S|-F)}$$

One probability we know is $Pr(S|F)$: Poisson showed that S is a logico-mathematical consequence of F, so $Pr(S|F)$ is 1. But what about the other probabilities—$Pr(F)$, $Pr(-F)$, $Pr(S|-F)$?

Suppose Fresnel's wave theory is not very probable, but not completely implausible; let $Pr(F)$ be 0.1. $Pr(-F)$ is now fixed as 0.9. Nobody expected to find a bright spot before Fresnel came up with his theory, so let's suppose it would be *really* unlikely that there would be a spot if the theory were false: $Pr(S|-F) = 0.0001$. Inserting these values and doing the arithmetic tells us that $Pr(F|S)$ is 0.99. The probability of Fresnel's theory has jumped from 0.1 to 0.99. Dramatic confirmation indeed! Bayesianism appears to provide a striking explanation of the evidential force of the bright spot (something appreciated by scientists at the time, and ever since).

We should be careful, however. Our probability values were generated quite casually. Suppose you think Fresnel's wave theory is rather less likely to be true—let $Pr(F)$ be 0.01. Moreover, as you reflect on $Pr(S|-F)$, you realize that you've never really considered the issue before, that you have no serious idea of all the potential rivals to Fresnel's wave theory, and that you are thus quite uncertain. You decide to set $Pr(S|-F)$ as 0.5. Now the value of $Pr(F|S)$ is 0.02. The probability of the theory has doubled as a result of the new evidence, but it is still very low. Moreover, if you suppose $Pr(S|-F)$ to be close to 1, you will discover that $Pr(F|S)$ doesn't shift far from $Pr(F)$.

So what probability *should* be assigned to Fresnel's theory in light of the evidence?

But serious questions arise about the Bayesian program. Scientists might talk vaguely about the probability of hypotheses, and about the increase of probability on the basis of evidence, but only in special circumstances do they venture precise estimates that could be used in serious calculation. Do they really *always* make such estimates secretly? And, if so, where do the probabilities come from?

Some occasions, as when we are drawing playing cards and when doctors are evaluating the results from medical tests, encourage precise claims about probability. We know that a deck has fifty-two cards, and suppose that any of them is equally likely to turn up; the doctors know, from past compilation of statistics, that the base rate of a disease in the population is so-and-so and the frequency of errors in the test such-and-such. When researchers entertain hypotheses about black holes or genes or tectonic motions, however, where can they find the information needed to make sensible judgments about probability? Bayesians' response to this worry might seem startling: It does not matter how the prior probabilities are set, they declare, for it turns out that the values assigned are much less important than they seem—indeed, they hardly matter at all. Bayesians therefore impose only very weak conditions on estimating prior probabilities: Any consistent hypothesis proposed is to be assigned a value strictly between zero and one.

The rationale for this lies in a mathematical theorem announcing the "washing out of the priors." If two investigators start with radically different prior probabilities, then, assuming they respond to the same evidence and behave like good Bayesians, their estimates of the probability of the hypothesis will converge over time as they continue the updating process. In the long run, their initial disagreement will be overwhelmed by the stream of evidence. As John Maynard Keynes famously remarked, however, "In the long run, we are all dead." It is worth pondering Keynes's observation a little. For, although the theorems about convergence bring good news, they also bring some disconcerting tidings. The good news is that our investigators will eventually concur. The down side is that, if we suppose that accumulating the findings takes time, and set bounds on the speed with which the results can be garnered, then, for any period we choose—however long—there will be different assignments of prior probabilities that do not yield significant convergence within that period. So there could be communities of scientists, all of whose members behaved in the Bayesian-recommended fashion, that would be unable to reach agreement on some matter of urgent importance within the time span allowed.

There is, however, a more fundamental difficulty. Successful use of Bayes' theorem demands the specification of *two* probabilities. Not only do you need to have a value of the prior probability of the hypothesis whose fortunes you are considering, but you must also make some estimate of the chance of obtaining the experimental finding if that hypothesis turns out to be false. There are some (rather uncommon) occasions on which there are grounds for that estimate. Typically, however, this will be another area in which individual investigators will have to make up their own minds.

Return to the example of the Poisson bright spot (named for a man who thought it absurd that there should be any such entity!). Initially, this looks like a wonderful example for Bayesianism—but the probabilities were introduced quite casually. The wave hypothesis, we said, initially had a low probability—but not too low. The claim that there was a bright spot, given that the wave hypothesis was false, received very low probability—and that was crucial. If we had supposed the chance of finding a bright spot was not so small, the impressive boost for the wave theory would have been greatly diminished. In the early nineteenth century, there were no sources for deriving an estimate of the chances of a bright spot. Perhaps, had they been asked, many scientists would have agreed with Poisson that it was very unlikely; perhaps some would have confessed that they had never thought about the matter, and would have concluded that any estimate they could make would be guesswork. Confirmation should not depend on anybody's guesses.

In our view, probabilities are useful tools for assessing the evidence in some scientific situations, namely in exactly those contexts in which the values can be responsibly assigned. Unfortunately, in many—probably most—contexts, there is no serious basis for judgments of probability. Consequently, Bayesianism fails as a general account of confirmation.

Theories

Theoretical science introduces many technical concepts that seem to apply to entities quite remote from observation: You can't be introduced to wave functions or covalent bonds or transcription factors in the way people show you petunias or koala bears. Yet, despite this remoteness from observation, it is commonly supposed that the language of theoretical science is unambiguous and precise, quite different from the cloudy notions employed in some discussions in the humanities and social sciences (*the social imaginary*, or *false consciousness*). People convinced of this difference have sometimes contended that the best way to help those immature fields that are struggling to achieve reliable knowledge is to require them to eschew the loose terminology that enables discussions to continue indefinitely without reaching resolution. A search for clarity sometimes inspires the *operational imperative*: Associate each of your theoretical concepts with a precise and definite criterion for application. During the twentieth century, many would-be reformers tried to "operationalize" areas of social science, proposing, for example, that attributions of psychological states—wants, intentions, emotions, and the like—should give way to clear stipulations about how subjects were disposed to behave under particular conditions. Instead of declaring that the rat wants the cheese, properly reformed psychologists talk only of the frequency with which an appropriate lever is depressed when cheese is in the offing.

Yet the natural sciences often use theoretical language with great success, even though they provide neither explicit definitions of their central terms nor operational specifications. How do they do it? Many champions of the analytic project have thought that careful attention to the *structure of scientific theories* would expose the justified ways in which the sciences use esoteric language, and highlight the differences between their procedures and those of the confused discussions of the humanities and social sciences.

Mathematical logic supplies a straightforward concept of a theory. For the logician, a theory is a set of statements (axioms) together with their logical consequences. That concept has clearly figured in the history of the natural sciences: Descartes and Newton tried to emulate Euclid in presenting definitions, postulates, axioms, and rules. By the twentieth century, it was plain that these efforts could be improved by using the new tools of formal (mathematical) logic: Even Euclid's attempt at axiomatization failed to make all the underlying assumptions clear. Scientifically sophisticated philosophers showed how to recast Euclidean geometry as a theory in the most rigorous logical sense, and they also aimed to do the same for other scientific theories—for Newtonian dynamics, the special theory of relativity, and the theory of evolution, in particular. In some of these examples, especially in those from physics, they were successful. This project inspired a general conception—the "received view" of scientific theories—that promised to explain the proper use of technical language. Scientific theories are axiomatic systems with vocabulary (apart from logical and mathematical terms) that divides into two parts. The *observation language* consists of terms that can be learned in application to items in the observable parts of nature. The *theoretical language* consists of those nonlogical terms that cannot be so learned. The axioms of the theory divide into three main types: First, there are *theoretical postulates*, whose only nonlogical vocabulary is theoretical; second, there are *correspondence rules* that contain both theoretical and observational vocabulary; third, there are *empirical laws*, whose only nonlogical vocabulary is observational. The guiding thought behind the received view was that the theoretical vocabulary is given meaning because the correspondence rules link its technical terms with observable phenomena. Unlike the writings of woolly intellectuals, who chatter on about transference and the *Zeitgeist,* physicists formulate principles that anchor their esoteric notions to nature. The term *electron,* for example, is partially interpreted by adopting a correspondence rule that identifies the kinds of tracks electrons will make in cloud chambers. The term *gene* gains meaning from similar principles that tell us how combinations of genes in systems of mating will be manifested in distributions of observable properties among the progeny.

As in the case of confirmation, the approach to theories began with a plausible idea, and then became bogged down in a host of technical difficulties. One challenge emerges from logical theorems that appear to show that every theory has a rival with axioms couched purely in observational terms, yet with *observational consequences* (consequences that contain solely observational vocabulary) that are exactly those of the original. This suggests that theoretical vocabulary can be dispensed with entirely without any loss in the theory's ability to make empirical predictions—rather as the operationalizers hoped to do in the social sciences. If that is so, why should scientists introduce theoretical vocabulary at all?

In response to challenges like this, some philosophers (and scientists) have adopted *instrumentalism,* the thesis that the technical terms of science are, strictly speaking, meaningless—that theories are simply tools for the convenient prediction of observable phenomena. They hold that genes and electrons, like the heavenly spheres of Ptolemaic cosmology, are merely "useful fictions" that help scientists see patterns among the phenomena they observe. Others have maintained that *scientific realism* is the appropriate stance—that it is reasonable to think of theoretical terms as (at least partially) meaningful, and to conceive those terms as picking out real constituents of nature. Scientific realists believe that theories do more than merely predict, and that the real entities they introduce help us to understand why the observed phenomena occur.

To make progress with these issues, it is important to note a deep problem with the motivation for the received view (a difficulty first pointed out by Hilary Putnam). The view operates with a pair of distinctions. One, the distinction between observation terms and theoretical terms, applies to parts of language. The other, the distinction between observable things and unobservable things, applies to parts of nature. The motivation for the project began with the thought that observation terms are those that can be learned in application to observable things. It then assumes that terms learnable in application to observable things apply only to observable things, thus generating the central conundrum of how scientists introduce language that talks meaningfully about unobservable things (if such there be). The assumption supposes that there is a neat coordination of the two distinctions. Observation terms apply just to observable things; theoretical terms apply just to unobservable things. But as Putnam noted, this is incorrect. On the one hand, there are perfectly ordinary theoretical terms, terms that you cannot acquire from observation alone but only through learning a theory, that apply to observables; for example, *electron microscope.* More important, there are terms you can learn in application to observable things that can then be employed to characterize unobservables: As Putnam wryly remarked, "Even a child can talk about people too little to see."

What does this imply for the introduction of scientific language? Consider the term *part.* It is plausible to suppose that you can learn this term by being shown middle-sized objects and their constituents (chairs and their legs, for one obvious example). Once you have acquired the vocabulary, however, you can make more ambitious uses, even thinking about objects that have no parts but are themselves parts of all other objects. Conceptions of that sort pervade discussions of atoms, from ancient writers to the nineteenth-century proponents of atomism (e.g., Dalton).

There is an even deeper difficulty. The received view tied theories closely to language, effectively identifying a theory with a linguistic formulation. Reflection on some areas of science can inspire an alternative approach, one that takes theories to be families of *models*. These models may be mathematical structures or they may be more concrete entities. So, for example, a dynamical theory might be presented by setting up a phase space and allowable trajectories through it (models as mathematical structures), or the Watson–Crick theory of the structure of DNA might be identified with the structure Watson and Crick actually built from wire and pieces of sheet metal (and the many more elegant and expensive versions that have been constructed since).

The alternative enjoys greater success in coping with parts of science often neglected by the received view. The axiomatizations of physics, inspired by enthusiasm for the recent development of mathematical logic, were often valuable in showing just which theoretical assumptions were needed for different purposes. Attempts to force the same structure on Darwinian evolutionary biology, however, were almost comic. Perhaps the most important part of biology emerged as a cumbersome collection of trivialities, utterly divorced from everyday scientific practice. Here the view of theories as families of models sheds far more light, and can be usefully applied to the debates that arise in evolutionary research.

This point should not, however, encourage a declaration of victory in philosophical debate about what scientific theories really are. For the question of the true nature of scientific theories is a bad one. The analytic project has sometimes been drawn into that question for a perfectly understandable reason: It wanted to expose the differences between responsible scientific language and the less substantial esotericism of other discussions. That goal was never attained, but in pursuing it philosophers sometimes offered valuable accounts of particular areas of scientific work. Their axiomatic presentations of the special theory of relativity, for example, made it apparent just which assumptions were required to support particular relativistic hypotheses. Analyzing a theory as a family of models can also serve particular purposes, illuminating a debate among practitioners about the best way of describing a particular range of phenomena—this has been valuable, for example, in discussions about the "levels of selection" (Does natural selection always select for traits of individual organisms or for traits of genes or does it also select, at least sometimes, for traits of groups?). Perhaps it is best to view scientific theories as a motley mix, deployed in research practices in very different ways. Combining that conclusion with Putnam's insights into the ways scientific language can be understood, the motivation for the analytic project's investigation of scientific theories evaporates.

Explanation

According to a traditional view, focused in an influential discussion by the French physiologist Claude Bernard, the goals of scientific inquiry are prediction, control, and understanding. The first two of these are practical: Scientific predictions prepare people for what will happen under various circumstances, so they can adjust their behavior accordingly; interventions are aimed at reaching outcomes people want to achieve. Yet the idea that the sciences are valuable only to the extent that they facilitate our practical plans seems crass and utilitarian. Isn't "pure science" worth having, even if there is no great use anyone can make of it? Like Bernard, many people have thought so, and they have introduced the third goal, the provision of understanding, to express their sense of the importance of pure knowledge of nature.

The sciences provide understanding by giving explanations. For the analytic project, explanation was itself something to be explained. Philosophers have agreed that the status of something as an explanation is not the result of some warm subjective feeling, a sense of "being at home" with the phenomena explained. On the contrary, many scientific explanations succeed by showing familiar events and states of affairs in a novel, sometimes even disconcerting, light: We come to understand, for example, that the night sky is not a blaze of light because of the expansion of the universe. What, then, goes on when a satisfactory explanation is provided?

One obvious proposal is that to explain something is to show how it came about by identifying the causes that produced it. In the early stages of the study of scientific explanation, philosophers resisted this idea. Influenced by Hume, they supposed that the concept of causation cannot be taken for granted. For Hume famously argued that our experiences of causal connection never reveal to us the necessary link between cause and effect: We may see the match light and the wick of the candle catch fire, but we have no insight into the intrinsic connection between the two (science would be far easier if we did!). So, Hume and his successors believed, our grasp on the notion of causation is exhausted by the patterns of succession we observe or infer. To the extent that we have a clear notion of causation at all, it results from our prior understanding of what kinds of earlier events explain later occurrences. Explanation explains causation, not the other way around.

The analytic project's systematic study of explanation began with a series of seminal articles by C. G. Hempel. From the late 1940s to the 1960s, many philosophers of science agreed that Hempel's proposal, the *covering-law model of explanation,* had delivered a clear, precise, and correct account of explanation. According to that model, explanation proceeds by deriving a description of the event (or process or state of affairs) to be explained (the description is known as the *explanandum*) from a set of premises (collectively the *explanans*), one of which must be a law of nature. (Among the supposedly minor difficulties to be resolved was that of saying precisely what a law of nature is; we shall consider this task shortly.) So, for example, we might explain the fact that the length of the shadow cast by a flagpole is twenty feet, as follows:

1. The height of the flagpole is 20'.
2. The angle of elevation of the sun is 45°.
3. Light is propagated in straight lines.
So: 4. The length of the shadow is 20'.

Here 1 through 3 make up the explanans, and 4 is the explanandum. The explanation derives 4 from the stated premises, together with some principles of logic and mathematics that do not need to be stated; here the derivation is deductive. We achieve understanding in this instance because, after we have the explanation, we see the phenomenon (the length of the shadow) as expected in virtue of the *framework of natural laws*. That, in the covering law model, is the heart of explanation: not to provide a subjective feeling of at-homeness, but to permit us to see aspects of our world as *to be expected*—either as bound to occur, or at least highly likely—in terms of the underlying laws of nature. Explanation takes its place as a close parallel of the other broad aims of science: Scientists are able to explain or predict natural phenomena, and to control what happens by means of interventions with predictable outcomes, all by drawing inferences from underlying laws.

Hempel showed how to extend this approach to cope with examples of probabilistic and statistical explanation. Sometimes we explain what happens by citing features that render it probable; sometimes we explain statistical distributions by deriving them from more fundamental statistical laws. He showed how the model could reconstruct the explanation of less basic generalizations from more fundamental principles (as when we explain Kepler's laws by showing them to be consequences of Newtonian gravitational theory). He also responded to criticisms that the account could not be extended into domains—like history and anthropology—in which it is hard to state the laws that explanations are alleged to need (it is assumed that what historians and anthropologists typically provide are "explanation sketches," in which laws that cannot be stated exactly are presupposed).

One central difficulty with Hempel's approach to explanation was its reliance on the concept of a law of nature. Laws have typically been identified as unrestricted generalizations, universal claims that attribute a property to all members of a population. Yet it has long been evident that not all statements of the form "All As are Bs," even if true, would count as genuine laws. Hans Reichenbach invited his fellow philosophers to consider the two generalizations:

(a) All spheres of pure uranium are less than 1 km in diameter.
(b) All spheres of pure gold are less than 1 km in diameter.

Assume, as seems overwhelmingly likely, that both A and B are true. There is supposedly an important difference between them, in that the first carries a kind of necessity (*natural necessity*) that the second lacks. We can bring out the intended contrast, by imagining that people try to build spheres of diameter greater than 1 km using each of the two materials. Their uranium project would *have to* fail—a sphere of that size is *impossible*. Long before the required size was achieved, radioactive processes would occur that would halt the endeavor (and probably obliterate the builders). By contrast, no such obstacle would prevent the construction of the gold sphere.

Yet now we obviously have a new problem on our hands. What is this natural necessity? How do we recognize it? And how can we fit it into our philosophical understanding of science? Some people have suggested that the world is intrinsically divided up into types of things—so-called natural kinds. Thus, for example, chemical elements are natural kinds, as are biological species. It is then proposed that natural kinds have *essences,* underlying constitutions that make members of those kinds what they are and from which many of their properties flow. In the chemical case, for example, there are distinctive features of the atoms of each element: To be hydrogen (to take the simplest case) is to be made up of atoms, each of which consists of a single proton and a single electron. The laws about hydrogen derive from this atomic structure. It is necessary that hydrogen atoms have this structure, and the necessity is passed on to the derivative laws.

Even in the chemical case, there are grounds for worrying about this account: difficulties with ions and isomers, for example. Matters are far worse, however, when attention turns from chemistry to biology, for it is central to post-Darwinian biology that species consist of variable populations: There are no properties essential to *Drosophila melanogaster* (the famous fruit fly) or to *Homo sapiens* (see "The Biology of Race"). The thesis that all natural kinds come equipped with essential properties is consequently highly dubious. Or should we say instead that there are no laws in biology, and, as a result, no biological explanations? Would that be allowing the philosophical tail to wag the scientific dog?

ᴖ The Biology of Race

The two most obvious examples of natural kinds are the chemical elements and biological species. Philosophers who have thought that natural kinds are distinguished by essences, from which the properties of members of the kind flow, have been inspired by the chemical examples. At first sight, it does seem plausible to contend that carbon has an essence, and that this essence is constituted

(cont).

in part by its having atomic number 14, although the existence of ions and isomers generates complications for this proposal. On the other hand, ever since the mechanism of inheritance was discovered, ideas about some genetic property found in all organisms of a biological species have been suspect. Biologists recognize considerable genetic variety within populations of the same species.

The often-tragic history of thinking about human races has been dogged by essentialist ideas. Building on the thought that races have different biological essences, people have supposed that the essential properties of members of some races mark them out as inferior in particular respects. The superficial markers—skin color, hair texture, and the like—are often taken to indicate the presence of underlying biological traits, essential to the race, which determine important features of character and intellectual ability. Not only is this doctrine flawed because of its commitment to essentialism, but it is also at odds with research demonstrating intraracial variation in the pertinent features. Despite periodic attempts to argue that members of some races are genetically disposed to have lower intelligence, critical studies have shown the flaws in the supposed evidence.

Because the attempts to rank races have been so persistent, many anthropologists have recommended abandoning the notion of race entirely, seeing it as an unwelcome social construct. Biology, however, continues to deploy the concept of species, identifying species without supposing that they have essences: One prominent approach takes species to be collections of populations that would freely interbreed in nature and that would not freely interbreed with populations from different species; in the preferred terminology, species are *reproductively isolated* from one another. Many biologists also recognize different units within the same species, speaking of "subspecies" or "local races," and differentiating these subgroups by their *partial* reproductive isolation from one another. If those ideas are legitimate for other species—in classifying shrubs or snails, say—why can't they be applied to human beings as well?

The development of molecular biological techniques during the past decades has made it possible to analyze the frequencies with which various nonfunctional ("meaningless") DNA sequences occur in human populations, and, on this basis, to recognize how the gradients of free interbreeding have gone in the human past. In 2002, Noah Rosenberg, Marcus Feldman, and their associates showed how those techniques could be used to generate divisions of our species that showed the closeness or distance of biological relations. Although the authors were very careful to avoid the term *race*—they wrote of "genetic clusters"—reports in the press immediately characterized the research as rehabilitating the concept of race.

So are there races in nature? Rosenberg, Feldman, and their co-authors have shown that there are frequency differences in particular molecular sequences among human populations, but it does not follow that nature has drawn boundaries that science has to respect. Perhaps how things are assorted

(cont).

is up to us, a question of what we find salient and useful for our purposes. Mindful of the past abuses of racial notions, we might agree to investigate the differences within our species without making sharp divisions. Or it might be suggested that drawing boundaries is useful for demography, for medical recruitment of organs for transplant purposes, or even for forging a kind of solidarity that will help groups who have historically been oppressed obtain justice. An apparently scientific question is entangled with issues in ethics and in social theory. From our perspective, the future of the concept of race depends not only on the kinds of facts that population geneticists discover, but also on the goals that are properly set (see Chapter 6).

A survey of scientific discussions shows that talk of laws is extremely unsystematic. In many domains of science, practitioners prefer to talk of models rather than of laws or principles or rules. Major hypotheses of flourishing sciences are known to have exceptions ("DNA molecules are double helixes," "Genes are transcribed to make RNAs that are translated into proteins"). Perhaps philosophers should focus on those generalizations that are repeatedly used in solving the problems tackled by the various fields of science, recognizing them as a heterogeneous lot, many of them tacitly tolerating exceptions. There is little reason to believe that a systematic account of them is possible—or even necessary. Maybe the notion of a scientific law is simply the residue of an older conception, in which the universe was designed by a wise creator. Once it might have been important, not only to natural philosophy but to natural theology, to reconstruct the rulebook, but in a more secular world the concept of a law of nature could be allowed to lapse.

Even if philosophers succeed in giving an analysis of the notion of a law (and they continue to try), the Hempelian account of scientific explanation faces further major difficulties. The model permits derivations that are seemingly not explanatory to count as explanations. Consider the flagpole example offered earlier. Given the law that light is propagated in straight lines, we can indeed deduce, using trigonometry, the lengths of shadows from the heights of flagpoles. Equally, however, we can derive the heights of flagpoles from the lengths of the shadows they cast. If you exchange statements 1 and 4 in our example, the result remains a valid argument, and it still uses a law of nature to derive the conclusion. Hence, if the original meets Hempel's conditions, so too does the amended version. Yet we do not typically think that we can explain the heights of flagpoles by citing the lengths of their shadows. That sort of derivation seems to get things the wrong way around. (You might be tempted to propose, at this point, that flagpoles *cause* shadows, but that shadows do not *cause* flagpoles. The response is a natural one, but it would not appeal to anyone influenced by Hume's critique of causation.)

Examples like this suggest an asymmetric relation—a relation of explanatory dependence—that the covering law model fails to capture. More fundamental, however, are concerns about probabilistic explanation, for they strike at the underlying idea of explanations as arguments that make phenomena expectable as consequences of laws.

First consider a medical example concerning the fate of Friedrich Nietzsche. Nietzsche died eleven years after a breakdown brought on by paresis. Why did he have paresis? Because, in his youth, he contracted syphilis from a prostitute: Paresis, with its dreadful (and lethal) neurological decay, sometimes occurs when syphilis is untreated. That is at least an outline explanation of the tragic end of a great philosopher. Yet the account does not make Nietzsche's death probable. Nobody develops paresis without having contracted syphilis first, but the chances that untreated syphilis will lead to paresis are only about 15 percent. So, although the explanation makes the mode of death more probable, it does not show that it was to be expected, for the probability remains quite low.

Worse is to come. Consider the phenomenon of electron tunneling. If a beam of electrons is shot toward a barrier, then, according to the principles of quantum physics, most will be reflected back. Depending on the character of the barrier, however, there will be some small probability that an electron will pass ("tunnel") through. Imagine that you are observing an experiment of this sort, and the fluorescent screen beyond the barrier reveals that an electron has struck it. How do you explain that event? All that can be done is to use quantum physics to derive the probability, a probability that may be extremely low; no further explanation can be given for why this electron penetrated the barrier.

In our world, the improbable sometimes happens. Unless you suppose that unlikely events are simply inexplicable, the idea that we can only explain what happens where we can present an argument that makes it seem expectable must be inadequate. In ordinary life, we do try to explain what we take to be improbable, typically by describing its causal background—as with the example of Nietzsche's death. Nor do we suppose that the possibility of explanation is unrealized until we have achieved some underlying account that shows why things had to be the way they were, or why (at least) they would be likely to be so.

In the wake of a series of critiques of the covering law model of explanation, which developed the points of the preceding paragraphs as well as exposing other troubles, philosophers of science attempted to find alternative general accounts of explanation. Some abandoned the Hume-inspired taboo against appeals to causation, and proposed that explanation always exposes the causal details leading up to the event, process, or state of affairs to be explained. Proposals of this type, even if they come equipped with a satisfactory theory of causation, are vulnerable to the objection that theoretical explanations usually do not seem causal in character. Kepler's laws are explained by deriving them from Newtonian gravitational theory, but it does not seem right to suppose a causal connection here. In human populations, the sex ratio at birth is biased in favor of males, and this fact is explained by showing how evolution under natural selection favors a 1:1 sex ratio at sexual maturity, and noting the greater vulnerability of boys in early childhood—that explanation advances our understanding in a way that grinding out the causal details behind the individual births never could (see "Dr. Arbuthnot's 'Male Years'"). Other philosophers, more concerned with theoretical explanation, have suggested that science explains by unifying phenomena—a sentiment expressed by many practicing scientists in their "philosophical" moments—but, even when the notion of unification is analyzed, it is hard to suppose that all successful explanations fall under it. Sometimes our explanations embrace diversity: In medicine, for example, we can recognize a variety of different ways in which people can acquire a particular disease.

⋙ Dr. Arbuthnot's "Male Years"

In 1710, Dr. John Arbuthnot examined the records of births in London for the previous eighty-two years, and discovered that every year was "male"—by which he meant that each year contained a preponderance of male births. Calculating the probability that this sequence of "male years" would occur by chance, he showed that probability to be truly minute, and ascribed the phenomenon to divine providence.

How do you explain a sequence of eighty-two male years? If you think explaining requires identifying the causes that produced a sequence of events, it appears you should be interested in tracing the individual etiologies of all the London births. That account—the *Gory Details*, as we'll call it—is too long for us to write in this box, but we can give a recipe for it.

1. Start with the first birth of 1628. Go back to 1627, and the act of copulation that began the pregnancy. Trace the movements of the sperm to the fertilization of the egg. Derive the chromosomal character of the sperm that won the race. Follow the process of pregnancy and birth. You have now explained the sex of the first birth of 1628. Enter it, under M or F, in a tabulation.

2. Repeat this for the second birth of 1628, and continue through all the births of 1628. Now add up the totals in each column of your tabulation. You have now explained why 1628 was a "male year."

(cont).

3. Repeat the same procedure for 1629, 1630, and all the years up to 1709. You now have a complete causal explanation of why all these years were "male."

We suspect you will feel that following this recipe does not provide much insight into the surprising fact Arbuthnot discovered: Even if you knew the Gory Details, you would still wonder at the coincidence (and we suspect you would not invoke divine providence). Notice that knowing the Gory Details doesn't shed any light on what is expected in 1710 (which was also, as it turned out, a "male year").

A better explanation comes from evolutionary theory. R. A. Fisher, one of the founders of population genetics, also began *sex-ratio theory*. In many (but not all) organisms that reproduce sexually, including human beings, there is an equilibrium value of the sex ratio at sexual maturity of 1:1. This means that if the sex ratio deviates from 1:1, organisms with a tendency to produce the underrepresented sex will be favored by natural selection. If boys predominate and girls are relatively scarce, couples with a tendency to produce daughters will tend, on average, to have a larger number of grandchildren (each birth requires one male and one female, and because there are more sons than daughters, the probability of a son fathering a grandchild is less than that of a daughter giving birth to a grandchild). Natural selection keeps the sex ratio at sexual maturity close to 1:1.

So why, in the large population of a city, are years "male"? Because the population is big enough to make it highly improbable that the sex ratio at maturity deviates far from 1:1—and because the mortality of boys (at least for most of human history) is greater than the mortality of girls. If you are to have a 1:1 ratio at puberty, more boys have to be born to compensate for those who will be lost in the intervening years. Part of the explanation is that in our species males are (in this sense) the weaker sex.

Like the quest for a theory of confirmation and for an account of the structure of scientific theories, the search for "the structure of explanations" has not delivered a general account. As philosophers reflect on the diversity of scientific practice, it becomes ever clearer how varied the questions are that researchers seek to answer. Sometimes they want to know how some interesting thing is composed or constructed; sometimes the task is to understand how it is possible that something has come about; sometimes, as in the human sciences, they want to know what it is like to live in a particular way; and sometimes the goal is to recognize how a result is generated. All these different types of question coexist with the search for why things happen, which, on traditional views, is the core of scientific explanation. Answers to a variety of types of question form part of the "pure knowledge" we rightly value. Bernard's goal of providing understanding is a much more diverse and untidy category than it has been taken to be. The prospects for recognizing all the instances of understanding as variations on a single theme seem dim indeed.

Failures—and Successes

The analytic project sought general accounts of confirmation, theories, and explanation partly in the hope that those accounts would help resolve debates that occur in fields of the natural sciences, and, even more prominently, in areas of the social sciences and humanities. As we noted, it is plausible to expect that reflection on the most strikingly progressive parts of inquiry would advance human knowledge in less fortunate domains. Yet, as we have seen, in each of the three main areas, the search for generality failed.

That does not mean the quest was fruitless. Reflection on confirmation, on theory, and on explanation has exposed some characteristics of some sciences, supplying concepts and principles that, although not fully general, are useful tools for tackling particular debates in physics or biology, for example. A mixture of philosophical analysis, supplemented with serious attention to crucial details about some individual controversy, has enabled philosophers to take up special problems in areas of natural science—to work with natural scientists on questions in the foundations of quantum physics or on debates about the credentials of human sociobiology. The analytic project was valuable not because it attained its intended goal, but rather because its attempts to reach that goal furnished tools for more piecemeal types of philosophical work.

One moral that might be drawn from this conclusion is that philosophical study of science should be thoroughly naturalistic. Naturalists in philosophy insist that the methods philosophers use are those of good inquiry generally: There are no special powers that philosophers can exercise from their special armchairs. Instead, the task of philosophy is to look carefully at some domain (e.g., science), to form and test hypotheses, and to work at piecemeal improvement of the existing practice in that domain. Seen in this way, the move away from the generalizations of the analytic project to a more modest engagement with the various sciences is a healthy development.

Yet there is one obvious cost of this retreat: It gives up on explaining what makes science such a thoroughly good thing. The thesis that science is a thoroughly good thing was our starting point, and it is worth returning briefly to it as we close this chapter. For, given a central idea—common to the analytic project and much everyday thinking about science—there is something odd about this thesis, or at least its confident assertion. The sciences are supposed to represent our very best forms of knowledge. Does the thesis that they do, itself, belong to this distinguished club? Apparently not, for, if it did, there would be at least one claim about goodness—a value judgment—that could be scientifically established, and that is widely viewed as an impossibility. Yet if the thesis is not something we know with that special scientific firmness, why is it held with such confidence, and why did it ever inspire such valiant efforts to explain the distinguishing features of science?

Behind the analytic project lurks an interesting complex of assumptions. Although philosophers have rarely discussed it explicitly, almost all champions of the project have tacitly accepted the common view that science is, or at least should be, a value-free zone. One distinctive feature of scientific investigation is supposed to be its antipathy to wishful thinking: Scientists follow the evidence, without interference from their values and hopes. Another lesson drawn from the historical development of the sciences is that nature does not supply goals and purposes: The world disclosed by the physics, chemistry, and biology of the past few centuries is no longer suffused with value. Scientific investigation of nature cannot disclose to us what is valuable, or what we ought to want. It would seem to follow, then, that no such investigation can teach us that knowledge is good, and in particular that the especially firm knowledge delivered by the sciences is something to be prized. Without confidence in the special value of scientific knowledge, however, the motivation to discern the shared qualities that underlie such knowledge begins to waver. Why should we be concerned to help those "lesser" fields that never reach consensus?

Assumed answers to questions about value underlie the analytic project. Later chapters explore those questions in their own right.

Suggestions for Further Reading

The material of this chapter is extensively covered, not only in the major works of prominent philosophers of science, but also in almost all introductory books on the subject. For a classic—and lucid—introduction to the issues, written by the major architect of the analytic project, see Carl G. Hempel, *Philosophy of Natural Science* (Englewood Cliffs, NJ: Prentice-Hall, 1966). More recent texts for undergraduates examine later attempts to give general accounts of confirmation, theory, and explanation. See, for example, Peter Godfrey-Smith, *Theory and Reality* (Chicago: University of Chicago Press, 2003). A valuable introductory text with somewhat different emphases is Ian Hacking, *Representing and Intervening* (New York: Cambridge University Press, 1983). We strongly recommend looking at these books before consulting the more advanced and technical works listed next.

Almost all the chapters of Hempel's *Aspects of Scientific Explanation* (New York: Free Press, 1965) illuminate the motivations for the analytic project and reveal how it attempted to overcome the difficulties it encountered.

For discussion of the distinction between discovery and justification, see Hempel, *Philosophy of Natural Science,* Chapter 2. A highly sophisticated (and technical) defense of the possibility of systematic methods of discovery when proceeding from statistical data is provided by Peter Spirtes, Clark Glymour, and Richard Scheines, *Causation, Prediction and Search* (Cambridge, MA: MIT Press, 2000). Popper's approach to demarcating science is given in the title essay of *Conjectures and Refutations* (New York: Routledge, 1989), and, more technically, in *The Logic of Scientific Discovery* (London: Routledge, 2002). For valuable discussion, see Larry Laudan, "The Demise of the Demarcation Problem" and "Science at the Bar—Cause for Concern?," both available in Michael Ruse (Ed.), *But Is it Science?* (Amherst, NY: Prometheus Books, 1988).

Wesley Salmon provides an extremely clear and accessible introduction to many of the issues about scientific confirmation in *The Foundations of Scientific Inference* (Pittsburgh, PA: University of Pittsburgh Press, 1967). Another valuable introductory book is Brian Skyrms, *Choice and Chance* (Belmont, CA: Wadsworth, 2000). The most thorough discussion of the merits and difficulties of Bayesianism is given by John Earman in *Bayes or Bust?* (Cambridge, MA: MIT Press, 1992), although this is not for the mathematically uninitiated. Questions about the language to be used in formulating generalizations from partial evidence were focused by Nelson Goodman's "new riddle of induction," posed in his classic *Fact, Fiction, and Forecast* (Indianapolis, IN: Bobbs-Merrill, 1956); there is an extensive further literature (see, e.g., Douglas Stalker [Ed.], *Grue,* Chicago: Open Court, 1994).

The attempt to provide a general account of scientific theories as axiomatic systems is extensively reviewed by Frederick Suppe in his monograph-length introduction to *The Structure of Scientific Theories* (Urbana, IL: University of Illinois Press, 1977). A lucid account of many of the motivating questions is provided by Hempel in "The Theoretician's Dilemma" (in Hempel, *Aspects of Scientific Explanation*). Hilary Putnam argued for the difference between the observation–theoretic and observable–unobservable distinctions in "What Theories Are Not" (reprinted in *Mathematics, Matter, and Method: Philosophical Papers Volume I,* Cambridge, UK: Cambridge University Press, 1975). The approach to theories as families of models is elaborated by Bas van Fraassen in *The Scientific Image* (Oxford, UK: Oxford University Press, 1980) and, in an especially straightforward way, in Chapters 3 and 4 of Ronald Giere, *Explaining Science* (Chicago: University of Chicago Press, 1988). A different perspective on models in science is presented by Mary Hesse in *Models and Analogies in Science* (Notre Dame, IN: Notre Dame University Press, 1966).

The best account of the covering law model of scientific explanation is the title essay of Hempel's *Aspects of Scientific Explanation*. Wesley Salmon provides a thorough overview of the difficulties with the model, and of subsequent attempts to respond to them, in his *Four Decades of Scientific Explanation* (Minneapolis: University of Minnesota Press, 1992). Many of the principal articles from these decades are reprinted in Joseph Pitt (Ed.), *Theories of Explanation* (New York: Oxford University Press, 1988). Some recent attempts to develop theories of explanation are James Woodward, *Making Things Happen* (New York: Oxford University Press, 2003) and Michael Strevens, *Depth* (Cambridge, MA: Harvard University Press, 2008).

For difficulties with the conception of laws of nature see Nancy Cartwright, *How the Laws of Physics Lie* (Oxford, UK: Oxford University Press, 1983) and Bas van Fraassen, *Laws and Symmetry* (Oxford, UK: Oxford University Press, 1989). Marc Lange offers a careful attempt to respond to the problems and to develop a general account in *Natural Laws in Scientific Practice* (New York: Oxford University Press, 2000).

A good introduction to contemporary views about natural kinds is given in Joseph Laporte, *Natural Kinds and Conceptual Change* (Cambridge, UK: Cambridge University Press, 2004). For extensive discussion of the heterogeneity of kinds, see John Dupré, *The Disorder of Things* (Cambridge, MA: Harvard University Press, 1993). On the status of the concept of race, besides the suggestions already given for Chapter 1, there are a number of recent articles offering different perspectives: Robin Andreasen, "A New Perspective on the Race Debate," *British Journal for the Philosophy of Science*, 49, 1998, 199–225; Sally Haslanger, "A Social Constructionist Analysis of Race" (in Barbara Koenig et al. [Eds.], *Revisiting Race in a Genomic Age*); Michael Hardimon, "The Ordinary Concept of Race," *Journal of Philosophy*, 100, 2003, 437–55); and Philip Kitcher, "Does 'Race' Have a Future?," *Philosophy and Public Affairs*, 35, 2007, 293–317.

Banting's discovery of insulin is beautifully described in Michael Bliss, *The Discovery of Insulin* (Chicago: University of Chicago Press, 1982). Freud's case study of the Rat Man can be found in *Three Case Histories* (Freud's texts, edited by Philip Rieff; New York: Touchstone, 1996). A valuable commentary is Clark Glymour, "Freud, Kepler, and the Clinical Evidence" (in Richard Wollheim [Ed.], *Freud*, New York: Doubleday, 1974). Skyrms, *Choice and Chance,* provides a clear account of Bayes' theorem. For a careful philosophical history of the initial debate about Fresnel's wave theory, see John Worrall, "Fresnel, Poisson and the White Spot," in D. Gooding, T. Pinch and S. Schaffer (Eds.), *The Uses of Experiment* (Cambridge, UK: Cambridge University Press, 1989).

CHAPTER 3

The View from the Sciences

The Sciences on Their Own Terms

The analytic project, which has dominated the philosophy of science in the English-speaking world during the past seven or eight decades, began with readily understandable goals. It sought a precise and general account of scientific methods that would enable them to be extended into other areas of inquiry. In assessing the project, we have tried to understand how its claims about various aspects of science might be put to work. Our verdict is far from uniformly negative. Even when philosophers have tried to force all instances into some Procrustean form, they have often offered conceptions that are valuable in addressing some of the issues that arise in the sciences. Bayesian confirmation theory may sometimes be a resource in trying to determine whether a particular hypothesis gains support from the data; an axiomatization of some part of science may reveal that certain assumptions are not needed. Our evaluation has suggested throughout that the major achievements of twentieth- (and early twenty-first-) century philosophy of science lie in local analyses and explanations, rather than in the general accounts for which the analytic project yearned. Yet we also noted that some kinds of questions were neglected by the focus on seeking general accounts of confirmation, theory, and explanation. In this chapter we start to explore some of those neglected issues.

Our emphasis on the local successes of philosophical analysis is in line with a movement within philosophy of science of the past twenty-five or so years, a movement that has self-consciously emphasized the diversity of the sciences, and that has urged philosophers to let the various sciences speak for themselves. Instead of coming to science (viewed as singular) with preconceived epistemological or metaphysical aims, that movement tries to consider what recognizably philosophical questions arise within the different sciences, and how they might lead philosophers to rethink even some of their most cherished and fundamental assumptions. One example, which surfaced briefly in the discussion of scientific laws, is the confrontation of philosophical ideas about essences with the perspectives offered by evolutionary biology and stereochemistry.

In this chapter we consider some of the results of this philosophical movement, in which neglected questions emerge from taking the sciences on their own terms. We begin with a central theme of the movement: the diversity of the sciences.

The Ideal of Unified Science

One hitherto unconsidered aspect of the analytic project was its devotion to an ideal (or, as we shall see, to ideals) of unified science. Philosophers were not alone in feeling the attraction of this ideal. To this day, many scientists continue to stress the unity—or "consilience"—of science. What are they claiming, and why? Are they right?

We distinguish three different senses in which science can be taken as unified. One, perhaps the most basic, rests on a commitment to a thoroughly naturalist conception of the world. Naturalists (both philosophers and scientists) often present their views indirectly, by pointing out what they dislike. They have no truck with supernatural beings (deities and spirits), with disembodied minds, with dim metaphysical entities and processes (Platonic forms or processes of Pure Reason), or with any other species of "spooky stuff." More soberly, naturalism begins from our scientific picture of the world, resolving not to admit things that disciplined inquiry cannot countenance. Yet naturalists recognize that our present picture of the cosmos is almost certainly incomplete. In rapid succession we have come to know of atoms and electrons, of quarks and strings and branes, and we can hardly suppose that our intellectual adventures have come to an end. So the limit of what there is cannot be set by drawing up an inventory from the contemporary perspective, even if we are (as we should be) catholic, allowing not only physics, chemistry, and biology, but also the social sciences, history, critical studies in the humanities, and other disciplined fields of inquiry to make their contributions. Rather, what is taken to exist should be the totality of those entities that are properly introduced into our ever-evolving picture of nature by rigorous reasoning. Naturalists think we can be confident today about the atoms, the electrons, and the quarks, less sure about the strings and branes, and reasonably certain that deities and disembodied minds are not to be included.

This is a first thesis of the unity of science, centered on the conception of a thoroughly natural world, lacking "transcendent" beings. It is hardly uncontroversial, and religious scientists would surely dispute it. Consensus is reached, however, by treating naturalism as an appropriate methodological stance. Even though individual investigators might privately believe that there are more things in heaven and Earth than are dreamed of in the naturalistic worldview, they agree with the scientific commitment to attempting to explain phenomena in the terms naturalism sanctions. This consensus constitutes a very general view of the methodological unity of science, the credentials of which we consider later.

A more ambitious methodological claim, one that inspired the analytic project (see Chapter 2), would insist on a substantive "scientific method" shared by all scientific disciplines. Our survey of attempts to elaborate theories of confirmation suggests skepticism about this idea. There are plausible modes of reasoning—hypothetico-deductivism, for example—that not only pervade the sciences, but also crop up in other areas of academic inquiry, in detective investigation, and in the everyday deliberations of intelligent people. At that level, we can find methodological unity, but the particular ways of articulating these pervasive forms of reasoning depend on the specific problems under study and the past achievements of the fields in which those problems arise. If we choose, we can declare that archeology, biochemistry, oceanography, and plasma physics share a common method, but the serious canons of good reasoning in these fields—the kinds of things practitioners need to know to carry on their everyday research—are strikingly different. Archeologists need to know how to infer features of past human life from scratch patterns on artifacts (and how to identify the artifacts in the first place); biochemists need to understand how to infer molecular structures from diffraction patterns. Although methodology courses might begin with an emphasis on certain very general features shared by many sciences, they quickly diverge. Some aspiring scientists have to learn statistical techniques (different techniques in different sciences), and others must understand the powers and limitations of instruments and experimental techniques. Philosophers who take seriously the practice of the various sciences recognize this diversity—and find interesting problems about the justification of various approaches to validating hypotheses in the debates that exercise practitioners (as, e.g., in the dispute about the roles of observation in the wild and of controlled experiments as guides to understanding animal behavior; see "Understanding Animals").

✄ Understanding Animals

Many people who have pets, or who spend significant time with animals, feel the temptation to attribute thoughts and feelings to them—no less a scientist than Darwin recognized the tendency in himself. During the twentieth century, scientists studying animal behavior often sternly cautioned against the dangers of anthropomorphism (describing members of other species using terms from everyday human psychology). Some of them insisted on the

(cont).

operational imperative (see Chapter 2), and demanded that animal behavior be soberly characterized by talking about capacities and dispositions to perform bodily movements: Instead of "The pigeon sees the seeds and wants to eat them," say "In the presence of a pile of seeds the pigeon is disposed to move toward them, peck at them, and ingest them."

Philosophical recognition of the explanatory power of appeals to factors that cannot be directly observed, coupled with explanatory successes in some areas of science (particularly in linguistics and cognitive psychology) have recently replaced this puritanical approach with a more liberal attitude toward the understanding of animal behavior. Careful observations of monkeys and apes have inclined primatologists to suppose that some calls signal the presence of particular types of predators, that subordinate males sometimes sneak off with female "friends" behind a rock because they see that it enables them to escape the observation of a dominant rival, and even that chimpanzees recognize the intentions of other members of their troop and sometimes aim to help realize them, or—more ambitious yet—that some primates have a sense of fairness.

Although the attribution of psychological states to nonhuman animals is no longer taboo, there is a lively dispute about exactly when and what we are entitled to ascribe. Some students of animal behavior believe that the best evidence is obtained by sustained observations of their subjects under conditions that are close to those found in the wild, by either following animals in their native habitat or placing them in environments that closely resemble the natural conditions (but allow easier observation). Others suppose that individual— anecdotal—reports about particular episodes are worthless, because they cannot be replicated, and that more probative evidence comes from arranging carefully controlled experiments, in which the animals are regularly given opportunities to perform a particular task and allow for the collection of data that allow statistical analysis. Champions of observations in the wild (or in simulated "native habitats") suspect that such experiments distort the animals' behavioral repertoire, leading them sometimes not to perform actions they would otherwise have done, and sometimes to behave in ways that would not have occurred in the wild. Plainly, there are many refinements of the polar positions. Observation of animals can be recorded on videotape, so that many scientists can review what occurred, thus guarding against idiosyncratic biases. Observations can inspire experiments, and experiments can lead investigators to look for analogous instances under more natural conditions.

The methodology of animal behavior research is still being worked out, but two points are evident. First, methodology cannot simply be "read off" some general method that applies to all sciences across the board. Second, philosophy can play a role by clarifying concepts and developing comparisons among problems that arise in different areas of research, but it must be thoroughly informed by the specific details with which the researchers are confronted.

The most influential ideal of the unity of science is, however, not concerned with methodological questions, but with a vision of the structure of nature. It starts from the naturalist thought that there are no mysterious entities, so that living organisms are not distinguished from the inanimate world through their possession of some special vital substance or vital force, and human beings are not separated from the rest of animal nature by having "immaterial minds." If everything that exists is made up of physical constituents—organisms composed of cells, cells consisting of molecules, molecules built up from atoms, atoms organized aggregates of fundamental particles (to stop there)—then all are physical systems. Physics is then the fundamental science, one that should eventually supply us with a theory of everything.

The dominant version of the ideal of unified science conceives of a hierarchy of disciplines. At the bottom (or at the top if you prefer a different metaphor) is physics. Chemistry is simply the elaboration of that part of physics that studies the fundamental constituents of matter. Physics and chemistry together can be articulated to generate further sciences: Earth science, oceanography, and atmospheric science focus on the properties of particular physico-chemical systems; biochemistry is concerned with the structure and behavior of the molecules that interact in living things. From biochemistry, science can develop an account of organisms themselves, explaining the properties of individual cells, single-celled, and multicellular organisms. Extending further, it can yield accounts of the neurons, neural networks, nervous systems, and brains that some animals have. Psychology is an outgrowth of neurological understanding. From human psychology, science can extend further, to the study of the interactions between people, and ultimately to their economic, political, and social behavior. Eventually, we can trace a path all the way from physics to sociology.

Of course, nobody knows exactly how to give an account of society in the primitive terms of particle physics. The ideal does not claim that these explanations can be provided in practice, or even, given human limitations, that our species could ever actually achieve them. Instead, it is proposed that they are possible in principle. Many of those who honor the ideal of unity also think that science is advanced by bearing the ideal in mind, that fields progress most dramatically when they are able to forge connections with the domains immediately beneath them: Thus psychology is improved by explicitly relating hypotheses about mental functioning to the structure of our brains, and biology has made great strides by paying explicit attention to the biochemical structures and interactions among significant molecules.

It will be helpful to distinguish the reductionist strategy from the thesis of reductionism. *Reductionism* maintains that, in some sense to be explained, the higher level sciences in the hierarchy we have envisaged can be reduced to lower level (more basic) sciences, and that ultimately all can be reduced to physics. The *reductionist strategy* is more modest, recommending only that, when some objects of scientific study are composed of others, research on the higher-level objects might be aided by considering the constituents and what the scientific investigation of the constituents can tell us about them. You can accept the reductionist strategy without committing yourself to reductionism, and, indeed, you might use your concerns about potential failures of reductionism to make your own version of the reductionist strategy more precise. This may initially sound mysterious, but, as we shall see, it is a perfectly coherent idea.

Reductionism requires a sense to be given to reduction. The classic approach was pioneered by Ernest Nagel, within the framework of the axiomatic conception of scientific theories and the covering law model of scientific explanation. Nagel proposed that the core of the relation of reduction was logical deduction: To say that a theory can be reduced to another theory is, as a first approximation, to claim that the first theory can be deduced from the second. Chemistry is reducible to physics, in that all the laws that hold about chemical structures and reactions can be deduced from physical principles (specifically, by solving Schrödinger's equation for the systems of physical particles that make up the chemical entities—the atoms and molecules—involved). This example already brings out why the simple equation of reduction with deduction is only an approximation to a convincing account. Chemistry talks of atoms and molecules; quantum physics does not. The obvious remedy is to add premises that connect the language of quantum physics with the language of chemistry. We need *bridge principles* that link the concepts of one science to those of the other, specifically by identifying the objects discussed in the reduced theory as assemblages of the more fundamental objects of the reducing theory. So, Nagel proposed, a theory is reduced to another when all the principles of the first can be derived from the principles of the second, augmented by appropriate bridge principles.

Molecular genetics provides one of the most striking instances in which one domain of science has been brought into an illuminating relation with its immediate, more fundamental, neighbor. Can we view this science as arising from a reduction of an older science, classical genetics, to biochemistry? At first, this appears promising. Before the rise of molecular studies, geneticists knew that genes were segments of chromosomes, and they studied the patterns of transmission of genes, without any deeper knowledge of the constitution of the hereditary material. Nascent molecular biology revealed that genes are segments of nucleic acids (DNA in many organisms, RNA in a few), and, after Watson and Crick, we know the molecular structure of DNA. So it seems that the way is open to specify the genes as segments of molecules whose structure can be explicitly presented, and then to use principles of biochemistry to deduce the classical generalizations about the patterns of transmission.

Ignoring the minority of organisms whose genetic material is RNA, we seem to have a candidate bridge principle: Genes are segments of DNA. A moment's reflection, however, will bring home how insufficient this is. There are many segments of DNA, and only some of them are genes. Which ones? Recall that the point of bridge principles was to provide a structural characterization of the assemblages of more basic entities that constitute the less fundamental entities (molecules, for example, were to emerge as built up from elementary particles). There was excellent reason for that idea, for, in the end, we want to be able to use the principles of the more basic science, principles that apply to the constituent entities (elementary particles in the physics-to-chemistry case) to draw conclusions about the larger aggregates. Hence, it appears that we shall want a specification of what genes are that tells us the chemical structures of just those segments of DNA that count as genes.

Our understanding of DNA enables us to simplify the problem: We could specify which segments are genes by characterizing all and only the sequences of bases (the As, Cs, Gs, and Ts) that constitute genes. Thanks to the invention, refinement, and mechanization of gene sequencing, contemporary biologists know the sequences of a very large number of genes. Can we use this knowledge to generate a structural characterization? No. There are no recognizable structural features shared by the sequences of all and only genes. If there were, then the task of selecting the genes from the reams of genomic sequence data could go on more efficiently than it does! (In practice, scientists find the genes by using computer programs to look for *open reading frames*, regions of sequence with a significant gap between a sequence known to initiate transcription and a sequence known to stop transcription; once that has been done, biochemical assays explore whether pertinent products—RNAs or amino acid chains—can be found; this approach does not count as genes any of those regions that play a role in regulating transcription, regions that, on some accounts, would be identified as genes.)

It is possible to say what a gene is: a segment of DNA that is transcribed to give rise to an RNA, and, in many instances to an amino acid chain. That is a *functional characterization,* and, insofar as the fact that it is functional is important, it does not permit the application of biochemical principles. (Biochemistry has no laws that govern the behavior of things that are transcribed to form RNAs!) The characterization is not a bridge principle, but it is not entirely useless. It does, after all, permit the derivation of those generalizations about genes that depend on their being (not too large) segments of DNA. Nevertheless, we have a first antireductionist conclusion: Even in an apparently successful case, the linking of biology to chemistry, the bridge principles on which reduction depends are not available. We also have a diagnosis of the breakdown of reductionism. Sometimes the languages used in different areas of inquiry are at cross-purposes, in that one science uses structural concepts, whereas the other uses functional concepts. When that occurs, reductionism is likely to fail.

There is, however, a second problem that arises even if the first hurdle can be surmounted. If molecular genetics represents a reduction of classical genetics to biochemistry, then principles of classical genetics can be derived from laws of biochemistry. It is not easy to find generalizations in the practice of classical genetics that flourished just before the advent of molecular studies, but the following is a promising candidate:

> *Independent Assortment*: Genes that are located on chromosomes belonging to different pairs assort independently in the meiotic division that leads to the production of gametes.

Can we derive *Independent Assortment* from biochemical principles? Perhaps. For even though we cannot give a structural characterization of the genes, we might hope to deduce a stronger generalization—any DNA segments on chromosomes belonging to different pairs assort independently—from which *Independent Assortment* would follow as a special case. What would this deduction look like? It would be horrifically complicated, for it would have to consider all the possible biochemical structures and relations that could obtain at meiosis in all species, and show how all of these heterogeneous cases give rise to independent assortment. Nobody has the slightest idea of how to do that in practice, but, as reductionists will surely remind us, they claim only that it is possible in principle.

Yet there is an important question: Would anything be lost by replacing our normal ways of biological thinking with the allegedly "fundamental" derivation? How do we normally think about independent assortment? We envisage cell nuclei as containing chromosomes that line up in pairs just before meiosis (see "Pairing and Segregation"). Members of any pair can exchange genetic material (this is crossing over and recombination), but that does not affect chromosomes belonging to different pairs. At meiosis, one of the resultant members of each pair is transmitted to the gamete formed. Elementary considerations of probability tell us that the transmission of the segments on a chromosome from one pair is independent of the transmission of the segments on a chromosome from a different pair. As Figure 3.1 reveals, we are conceiving meiosis as a special sort of *functional process*: a process that works by bringing entities together in pairs and then randomly choosing one member from each pair. That is why the stronger generalization about the independent transmission of DNA segments holds, and it is why *Independent Assortment* obtains.

✑ Pairing and Segregation

Figure 3.1 shows two pairs of homologous chromosomes that have come together and recombined. On pair 1, there are two loci: one with alleles A and a, and one with alleles C and c. On pair 2, one locus is shown, with alleles B and b. Note that the transmission of alleles at the two loci on pair 1 is not independent. If a gamete (sperm or egg) receives the chromosome with A, it will also have the allele C; if it gets a, it will have c. By contrast, the transmission of A-alleles and B-alleles *is* independent. If the gamete formed receives A, the chance that it will have B is just the general probability that the chromosome containing B is selected. Under normal circumstances, when each member of each pair is equally likely to be passed on, the chances of each of the four combinations— AB, Ab, aB, ab—are equal; for any of these, the probability is 0.25.

Figure 3.1 and the last paragraph bring out the structure of the process, showing us why independent assortment occurs. What genes are made of is entirely irrelevant to the explanation. That explanation would work equally well if the genetic material were silicon or silk or Swiss cheese.

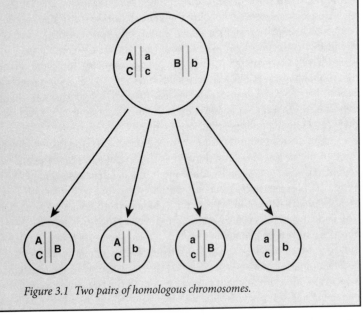

Figure 3.1 Two pairs of homologous chromosomes.

The last paragraph gives an explanation of a generalization about independent assortment. Would the explanation be deepened if we substituted the imagined (horrible) reductionist derivation? We think not. For, in its mass of detailed structural considerations, the envisaged deduction would not bring out the *functional unity* in all the cases. Independent transmission happens not because in case 1 this sequence of reactions takes place, in case 2 that, and in case 1,297,681,379,442,159 something completely different—it happens because all the instances, including some that we do not know about and some that might easily occur but do not, involve a simple process of pairing and selection. To understand the phenomenon, that is exactly what we need to know.

We conclude that the ideal of unified science that is committed to reductionism should be rejected. This is in no way to impugn the reductionist strategy. It is simply to recognize that a science like molecular biology is a synthesis, formed by using the structural analyses from biochemistry where they are illuminating, but not abandoning the functional language that is useful in dealing with biological systems. Indeed, once the point has been made, it is not hard to see that functional concepts are embedded within molecular biology at very basic levels. Molecular biologists talk happily of transcription factors and of the process of transcription, without being able to provide a full structural specification of either: Transcription is a process in which particular things get done and transcription factors are molecules that do those things. Perhaps not even molecular biology is reducible to biochemistry!

The ideal of the unity of science just considered built on the proposal that theories are axiomatic systems, and imagined the possibility of a physical supertheory from which the axioms of all "nonfundamental" sciences could be derived. As we saw in Chapter 2, there is a different way of conceiving theories, namely as sets of models, and adopting this approach can generate a different type of skepticism about the ideal.

The thought of a wonderfully unified science has appealed to the systematic ambitions of individual scientists for centuries. In the "Queries" he appended to his *Opticks,* Newton envisaged the possibility of a science that would extend his dynamics and his theory of gravitation to account for all the phenomena of nature, and some of his immediate successors were inspired by that vision. They believed that the three dynamical laws applied to all constituents of the cosmos, that Newton had characterized gravitation, and that the residual task of inquiry consisted in discovering all the other forces of nature.

There is a different way of thinking about Newton's achievement. He showed us how to model a particular sort of system, one in which gravity is the only force that matters. That is a very local result, one that can be applied here and there, so long as the perturbations from other forces can be ignored. For many falling bodies near the surface of the Earth, for example, we can model them and the Earth as a Newtonian gravitational system, and work out the details of the trajectory. Even if we restrict our attention to free fall, however, we cannot provide an analysis that will work for all cases.

Consider an example, originally offered by Otto Neurath and provocatively discussed by Nancy Cartwright. Imagine that you are standing on the top of a high tower on a windy day. You take out a twenty-dollar bill to reimburse a friend who has paid the entrance fee to the tower, and the wind whips it away. Where will it end up?

A natural reaction is to declare that this situation is governed by the principles of Newtonian dynamics. Unfortunately, nobody knows all the forces that are acting, so that the trajectory of the bill cannot be calculated. You might insist, however, that there *is* some unknown force function that applies on this occasion, and that, if only that function were known, the flight of the money would be derivable. The laws of Newtonian dynamics are taken to apply always and everywhere, even though we often do not know exactly how to work out the details. Cartwright invites us to consider the scenario differently. There are parts of the world that are orderly from a Newtonian point of view, either because approximations to Newtonian gravitational systems arise naturally (e.g., the motions of the planets in the solar system, the fall of heavy objects in the absence of perturbing forces, such as high winds and electromagnetic fields) or because human beings arrange them. Scientific inquiry focuses on these pockets of order, natural and man-made, so that *parts* of nature can be rendered predictable and controllable. By what right, however, do we suppose that there are ways of making aspects of our world that we have not yet tamed subject to the techniques that have been efficacious? Cartwright suggests that the idea that every dynamical process could be understood as a Newtonian system is an article of "fundamentalist faith."

You might reply that scientists have investigated motion under conditions of atmospheric turbulence. They build wind tunnels and explore what happens in them. On this basis, they are able to show that some effects of the wind can be understood within the Newtonian framework. Surely this provides a basis for generalizing, for thinking that a Newtonian treatment of the falling bill is in principle possible. Perhaps. Yet we do not know enough about the ways winds actually swirl in nature to be confident that the wind tunnel scenarios are reasonable modes of approximation. Why is it necessary to take a stand here? In practice, situations like this defeat our abilities to calculate, and it is not obvious that much is gained by declaring boldly that a computation we cannot hope to achieve is available in principle. In fact, as Neurath originally pointed out, we do not need to work out the trajectory of the bill to arrive at some answer to the original question. Low-budget economic modeling predicts that the bill will end up in a particular type of place—to wit, a human being's pocket.

The ideal of unified science supposes that there are universal laws of nature that apply always and everywhere, and that these are the fundamental principles from which all truths about nature flow. That ideal was already called into question by the discussion of molecular genetics, but Cartwright offers a significant new way of abandoning it. She sees science as a thoroughly human activity, in which individuals and communities focus on particular aspects of nature—aspects that are particularly important for them in light of their evolving needs—and try to find ways of making these parts of nature tractable. Sometimes they intervene, building special systems (often inside laboratories) that are shielded from the effects of potentially disruptive forces; sometimes they leave the world much as it is, attempting to construct models that will reveal to us what to expect. The search is not for overriding generalizations, but for methods of prediction and intervention that are well adapted to our purposes. As the sciences undertake that search, they offer a patchwork of orderly places, subsumed under a heterogeneous class of models. They give us a "dappled world."

Cartwright's vision is far more accurate as a depiction of the practice of the various sciences today than is the ambitious conception of inquiry as seeking, and uncovering, universal laws of nature. As her studies reveal, even within physics (the home base of the "universal laws"), much of the work that is pursued consists in building models of particularly significant kinds of systems. Generality is, of course, a good thing where we can achieve it, but there are rival aims for our inquiries—precision and accuracy, to name two obvious ones. Sometimes it is more important to have an exact account of the behavior of some system, or group of systems—one that can be relied on in our practical decisions—than to offer a more general model that would have to sacrifice either precision or accuracy. The purposes of scientific investigation can better be served by giving up any overriding commitment to the ideal of unified science.

The Ineradicability of Causation

One of the most striking aspects of the analytic project in its early phases was the attitude toward causation it demanded. As an empiricist, Hume had argued that causal relations—if they exist at all—are forever beyond the reach of empirical inquiry. In his capacity as a naturalist, though, he acknowledged that causal reasoning is fundamental to our natural way of learning about and dealing with the world, and went so far as to lay out a set of rules for drawing appropriate causal conclusions from empirical data. Many philosophers contributing to the analytic project, by contrast, took Hume's conclusions on the empirical inaccessibility of causal relations as grounds for scientific reform, proposing that mature sciences restrict themselves to making claims about mathematical relations among measurable properties—claims that can be empirically confirmed. They dismissed any appeal to causation as a pernicious intrusion of metaphysics into science, a relic of a bygone age that survives (Bertrand Russell remarked), like the monarchy, only because it is (wrongly!) supposed to do no harm.

Recommendations of this sort consorted well with attitudes and practices in many sciences of the early twentieth century. Physics—which philosophers accorded a special status both as the best exemplar of mature mathematical science and as the fundamental science to which others were to be reduced—seemed to some to have little need for explicit talk of causation. The hypotheses and laws of physics are not stated in terms of asymmetric cause–effect relations, but as mathematical equations that permit inference symmetrically in either direction. The new quantum mechanics, in particular, had some startling implications for the applicability of everyday causal concepts to the microphysical world. Most physicists thought that quantum mechanics showed that relations among events are fundamentally probabilistic, implying that the deterministic conception of causation could not be maintained, and quantum theory also predicted dependencies between the properties of particles arbitrarily far from one another that could not be handled in conventional causal terms. In the social and behavioral sciences, meanwhile, the early twentieth century had seen the elaboration of new statistical methods under the guidance of thinkers who largely shared the view of causation as outmoded: Karl Pearson, for example, developed the concept of *correlation* precisely as an alternative to the murkily metaphysical notion of causal relation. Generations of statistical modelers were trained to follow the dictum "Never infer causation!" and the sciences that depended on statistical methods—those in which controlled experiments were difficult or impossible to arrange—likewise clung to the safer language of correlation and association.

At the same time, though, both natural and social scientists struggled with the problem of how to learn about causal relations in different contexts, for (as was repeatedly pointed out) what we want science to tell us, often, is precisely about causes: what the effects of our actions or inactions will be; which factors are important contributors to outcomes that we desire or wish to avoid. Epidemiologists, for example, faced the challenge of distinguishing causation from mere correlation without recourse to experiment. After a few early skirmishes, nobody disputed that smoking was *correlated* with lung cancer. But did smoking *cause* cancer? (See "Smoking and Lung Cancer.") In 1965, leading epidemiologist Austin Bradford Hill offered his colleagues a set of criteria for distinguishing between causal relations and non-causal correlations in cases like this one, criteria remarkably similar to the rules Hume had laid out more than 200 years before. Similar efforts were made in other areas, and continue today, responding to the diverse problems that appear in different scientific contexts.

ᴐ Smoking and Lung Cancer

In the early twentieth century, lung cancer was a rare disease, so uncommon in fact that doctors at teaching hospitals would sometimes call medical students together to observe one of the few cases they were likely to see. During the early decades of the twentieth century, mass production of cigarettes coupled with successful advertising greatly increased the percentage of smokers in the population, and, by the 1930s, it was apparent that deaths from the disease had risen dramatically in frequency. Starting in the 1950s, epidemiologists began to amass evidence for extensive correlations between smoking and lung cancer.

The news they brought was not popular, and it inspired the tobacco companies to begin a vigorous campaign to defend the harmlessness of their products. Their central argument was that the mere fact of correlation did not demonstrate any causal link between smoking and lung cancer. That line of reasoning could be developed in a number of ways: by pointing out that statistical data also showed correlations between smoking and other diseases (which seemed quite remote from any possible physiological effect of smoking), by arguing that no mechanism for producing tumors had been given, or by claiming that the correlation might come about through the linkage to some previously unsuspected causal factor. In elaborating these defenses, the tobacco industry was able to recruit some prominent statisticians.

From the early 1950s to the late 1960s, medical researchers responded to this campaign by refining the methods of epidemiology. They produced ever more fine-grained studies that compared groups of very similar people who differed only in their propensities for smoking, and they differentiated levels of tobacco use and types of smokers (those who inhaled deeply vs. those who did not, those who used filters vs. those who did not). The confirmation of correlations between smoking and a variety of health problems gradually turned back the initial worry that the correlational studies had proved too much. Experiments on the effects of tar on the lungs of mice began to suggest a possible mechanism, and these were further buttressed by evidence from pathology, as the lungs of those who died from lung cancer were analyzed.

Perhaps the most creative defense offered by the tobacco companies was devised by the statistician and geneticist R. A. Fisher. An avid smoker himself, Fisher proposed that there was an underlying genotype with two distinct effects: People with this genotype have a strong desire to smoke, and the genotype also confers a propensity to develop lung cancer. If you are one of the unlucky ones with the genotype, you cannot improve your health by refraining from smoking—for the genetic propensity to acquire lung cancer will be there anyway. You might just as well go ahead and enjoy yourself.

(cont).

Fisher died before this inventive idea (for which there was no evidence) was strongly disconfirmed. Detailed epidemiological research on rates of lung cancer among smokers showed that the distribution of the supposed genotype among some populations would be quite at odds with the actual patterns of inheritance. Since the late 1960s, the connection between smoking and lung cancer has been part of medical orthodoxy, although tobacco companies continue to fight a rearguard action against the medical establishment.

A side benefit of the controversy has been the elaboration of epidemiological methods and statistical techniques that have been rigorously applied to understand the distributions of diseases and their causes.

In medical contexts, double-blind randomized experiments have come to be regarded as the best means of determining which are the causal relations. The special status of double-blind clinical trials is thus one crucial issue (especially because some treatments are very difficult to accommodate to double-blind experimental designs); another is what qualifies as a legitimate experimental control. In historical sciences such as paleontology or parts of cosmology, scientists need to find ways of establishing causal histories even though experimental intervention is impossible and the chain of events is unique, so no generalization over a class of like cases can be made. In climate science, the special challenges of tracing causal relations in hugely complex systems with multiple feedback loops lead to questions about the use of computer simulations for identifying causes and the correct treatment of low-probability extreme outcomes in predicting the effects of policy choices. In evolutionary biology, scientists and philosophers debate whether the effects organisms have on their own selective environments constitute a causal contribution to evolutionary change that is separate from natural selection. These debates, and many others like them, have proceeded despite the general doubts about causation that philosophers and statisticians have long expressed. That mainstream philosophy of science was, for several decades, almost completely irrelevant to these important debates within the sciences is a remarkable reflection on the priorities that the champions of the analytic project often embraced.

That time is over now, however. A sea change has taken place in the last few decades in the attitude of philosophers to causal talk. Rather suddenly, the fashion for empiricist austerity about causes has passed, and philosophers find themselves involved in a wide array of debates about scientific knowledge of causal relations. How did so radical a change come about?

Part of the story certainly is that, as philosophers began to be interested in naturalism and the promise of particular scientific approaches for addressing philosophical problems, and as their attention to the sciences spread far beyond mathematical physics, they became more willing to accept the kinds of causal talk employed in many scientific fields. But to do this without simply abandoning philosophy in favor of psychology, they needed a principled response to Hume's skeptical argument about causation. This need was filled by the introduction—or reintroduction—of just such a response. Soon after Hume's argument was originally published, Immanuel Kant had responded with a defense of the objectivity of causal relations that came to be called "Kant's answer to Hume." New arguments closely akin to Kant's played a crucial role in transforming philosophical attitudes to causation in the second half of the twentieth century.

Kant had seen the threat Hume's argument on causation posed to natural science, and his response was explicitly intended to rescue science (particularly Newtonian physics) from that threat. His basic idea was that Hume's story about how we get our knowledge of causation puts things the wrong way around. Hume thought that we begin with experiences of objects and events (which are just changes in the properties of objects), and that when we experience a pattern among events—that one type of event is always followed by another—we form the expectation that the pattern will continue, and so conclude that events of the first type *cause* those of the second. We then form our general conception of cause–effect relations on the basis of the various particular instances of this sort that we encounter. The problem, of course, is that the causal conclusions we draw are not justified by the experiences that give rise to them. Kant argued instead that it is impossible to have experience of objects and events in the first place without presupposing that they are causally related to one another. What does this mean? Think about how we see the world around us. What appears in our visual field is constantly changing, but we do not think all these changes correspond to events; some are just changes in our own perspective or the direction of our gaze, for example. So how can we pick out, from the constant flux in what we see, the objects and events that make up the world we experience? Kant argued that the only way we can do this is by using the concept of causation: To identify a sequence of appearances as an event is just to identify it as one in which something determined the appearances to appear in that particular order; that is, to identify it as the effect of some cause.

This response to Hume's views on causation does not help directly with the problem of how to confirm any particular causal hypotheses, but it gives reason to think causal hypotheses are not inherently unconfirmable, irrational, or nonsensical. Hypotheses about particular cause-and-effect relations seem instead to have the same status as any other empirical knowledge claims: They are uncertain and acceptance of them is always revisable and vulnerable to skeptical challenges, but reasonable efforts can be made to clarify how we might distinguish hypotheses that are well-supported by evidence from those that are not.

Kant's answer to Hume was revived, in part because of the increasing recognition of the role of causal concepts in many areas of science and in part because of developments within philosophy. Wilfrid Sellars argued that to apply even an ordinary empirical concept to something we experience is to commit ourselves to a set of implications about what other concepts will be able to be applied to it, under what conditions. To identify an object as having a particular property is thus to make a set of presumptions about what changes it would undergo under what conditions—i.e. to which causal laws it is subject. Hume had said that we can only observe the pattern that seeing fire (at close range) is often followed by a painful burn; we can never witness the causal link between the two experiences. Sellars says instead that the link is already there in the very concept of fire: that to identify something as *fire* carries implications about the conditions under which it will cause burns. Of course the presuppositions we make in using any particular concept may be mistaken, but we can revise our beliefs about causes in light of experience just as we learn about any other empirical matters. This means we can (reasonably) look for the patterns that permit inference (from cause to effect) and intervention (to achieve or prevent effects by acting on their causes)—which is what scientists in many fields have been doing all along.

Against the Supernatural

The return of causal thinking has helped bring our philosophical understanding of science closer to the practice of many kinds of scientists. But not all causes are equally welcome. We have noted that a commitment to a modest methodological naturalism grants some unity to the disparate sciences. According to this commitment, certain kinds of causal factors—"transcendent" or "supernatural" entities—are not to be invoked in the practice of scientific inquiry. An obvious and important philosophical question (especially for philosophers who aim to extend the naturalist attitude into philosophy itself) is whether this constraint on inquiry is a form of materialist dogmatism from which the sciences should be freed.

Supernatural beings were not always outside the purview of science. Great early modern thinkers, who were more or less orthodox Christians, thought of their work as directed toward fathoming the purposes of the almighty creator. By reconstructing the motions of the heavenly bodies, they would "think God's thoughts after him." Pious eighteenth-century investigators amplified the idea, suggesting that God had provided mankind with two books, the book of revealed scriptures and the "book of Nature"—the natural world. For those with the skill to read it, the second offered insight into the divine purposes. So, when English country parsons dutifully collected the flora and fauna of their parishes, and wrote up their descriptions as works of natural history, they were not neglecting their pastoral duties, but contributing to the proper celebration of God's providence, manifested in his concern for and intelligent design of his creatures.

Methodological naturalism rejects their attempt to connect empirical findings to claims about supernatural entities, but it cannot do so on the basis that the very concept of science debars transcendent causes. Contemporary science grows out of forms of inquiry that acknowledged no such excommunication. If methodological naturalism is to be something other than dogmatic dismissal of what appear to be logical possibilities, it must rest on reasons for giving up these older modes of scientific practice. So we should ask why scientific appeals to the creator died out.

Part of the answer lies in the increasing ability of the sciences to explain various aspects of nature without the invocation of divine action that had previously been popular. For the botanizing clergymen of the eighteenth century it was hard to envisage any other explanation of the "exquisite fit" between organisms and environment than the appeal to the wisdom of the creator who had individually formed the various organic types. From the early nineteenth century on, this position became increasingly untenable, first as researchers learned that the history of life on Earth was far longer than they had thought, and that the history had shown a succession of very different life forms. Then, in 1859, in the *Origin of Species,* Darwin proposed a radically different approach to the exquisite fit. Populations of organisms vary. There is relentless competition among organisms for leaving progeny in subsequent generations. Some of the variation contributes to an organism's ability to succeed in this competition, and is also heritable. Successful variants pass on their characteristics to their offspring, and so the traits underlying the success increase in frequency. Out of this process come the properties that looked initially as if they had been specially designed.

That, however, is only part of the story. It might have turned out that for an identifiable class of traits—showing especially exquisite fit—there was no possibility of providing the Darwinian style of explanation. Moreover, intensive study of the natural world might show that there are no instances of poor design in nature. If this pair of results had emerged, devout scientists would have been able to claim that the organic world is the expression of a wise deity, who has sometimes achieved his ends by the indirect route of using natural selection, and sometimes acted directly. There is a worldwide consensus among biologists today that we know of no instances in which Darwinian accounts are doomed to failure, and that the organic world is overflowing with instances of poor design (see "Bad Design Ideas"). Although there have been repeated attempts to attack the first part of this consensus, the second claim is indisputable.

That alone suffices to undermine the continued appeal to supernatural causes. For imagine (contrary to what one should reasonably believe) that there are some adaptations too exquisite for us to explain by means of contemporary Darwinian theory. To invoke a creator to explain the presence of these traits would raise the immediate question of why the poor designs persist. The supernatural powers allegedly available to do some things have not done others of the same generic type. Why? According to the canons of inquiry adopted by the devout investigators of the eighteenth century, and by their equally pious successors who regretfully abandoned the appeal to divine action, this question would have to be answered by an account of the systematic action of a wise creator. Faced with the extraordinary profusion of inferior designs, from the masses of pseudogenes and potentially troublesome repetitive DNA to the botched construction of organs and limbs, no wise scientist would rise to the attempt. Hence, given what we know about the organic world, the invocation of supernatural causes would stop the conversation by introducing a mystery about which nothing further can be said.

∽ Bad Design Ideas

Much pre-Darwinian biology celebrated the "exquisite design" of organisms. During the past 150 years, however, biologists have explored the many ways in which the structures found in plants and animals are, from an engineering point of view, extraordinarily unintelligent. Darwinian evolutionary theory often emphasizes the point, and its champions frequently claim that any creator who directly designed the organic world would fail to graduate from any reputable engineering program.

A famous example is the panda's "thumb." Stephen Jay Gould recorded his surprise at observing pandas at the zoo, clumsily holding bamboo between the "thumb" and the other digits: Pandas are not supposed to have opposable thumbs. Indeed, they do not, as Gould discovered when he counted five digits on one side and the "thumb" on the other. What looks like a thumb is an extended wristbone, which protrudes as an inefficient device that enables the panda to hold (and enjoy) its distinctive vegetarian diet.

If you were designing a mammalian body from scratch, you would surely try to avoid the complications that arise from the close packing of our reproductive and urinary systems. As mammals develop, the tubes have to be extended to enable each system to perform its function, and, in the two

(cont).

major types of mammals—marsupials and placentals—one sex suffers in consequence. Among the marsupials (like kangaroos), the awkward looping of tubes occurs among the females. In placental mammals (like human beings), the males experience the tangles. Either way, one sex is burdened with hernias waiting to happen.

As molecular biology has amassed ever more detailed knowledge of the genomes of various species, it has become ever clearer that genomes of multicellular organisms are full of the residues of sequences that were once functional genes, but now play no role in generating proteins. Worse still, there are long repetitive sequences that can increase beyond the point of merely being useless. If the repeats become too long, they can disrupt normal functioning and cause severe forms of developmental disruption and dementia: Fragile-X syndrome and Huntington's disease are well-known instances.

The human body is a mass of compromises, forced on us because we are locked in to the ways in which our ancestors solved their problems (problems often quite different from our own). We use our mouths and throats to take in both food and air—and anyone who has had a serious episode of choking knows about the problematic design involved. Our knees are modified versions of joints that were once shaped to cope with different ways of moving around, and, as people live longer, we come to feel the limits of the compromises the knee embodies. Perhaps the most dramatic bit of poor human design attends every human birth. Our erect posture sets limits on the width of hips; our large brains demand a broad channel for the baby's exit into the world. Human infants are born early: Were they to mature longer in the womb, the way out would be even riskier than it is. Birth is thus a fragile compromise. The baby comes out unfinished, set for a long period of dependence, and, even so, the head size stretches the mother. The consequences are not only painful, but also dangerous to both of those involved.

The examples given are only a very tiny sample from the huge number of bad design ideas contemporary biology now recognizes.

That verdict brings out an important motivation for methodological naturalism, one sufficiently powerful to recruit even religious scientists. As we look back on the history of the sciences, it is apparent how much has been achieved by *not* declaring that certain kinds of puzzling phenomena are intrinsically mysterious. Many things that have been discovered (often with much ingenuity and struggle) would have remained hidden, if scientists had been content to invoke a supernatural entity who acts in unfathomable ways. With that in mind, it is an entirely reasonable attitude to propose that the conversation should not be stopped. Methodological naturalists should, however, be prepared to revise their stance under certain conditions. If there were a serious hypothesis about supernatural causes that came to terms with the data we have, explaining how those causes operate and how they fail to amend the faulty designs of the natural world, then there would be grounds for attention and scrutiny. At the moment, the enterprise of producing any such hypothesis seems to have no takers.

Many religious scientists support methodological naturalism for these reasons. They suppose, on other grounds, that our world has been created by a wise God. Some of them think that it is a mark of the divine foresight for the creator to have set up the cosmos at the beginning, and then let it proceed according to the laws he had prescribed to nature, without further intervention. It is easy to see how scientists of this sort can adopt the naturalistic stance. Others hold that God has acted on some occasions since the original creation. They recognize, however, that the modes of divine operation are opaque to our human view, and that attributing various effects to the interventions of an unfathomable deity blocks avenues of potential research. So, although they might think that there will ultimately be limits to what scientific research can explain, they admit their ignorance of where exactly those limits are, and acquiesce in methodological naturalism.

Making Sense of Ourselves

We conclude with a brief look at the implications of a scientific worldview for our understanding of ourselves. As the tools of science are brought to bear more effectively on human beings—and on the workings of human inquiry itself—some special concerns arise. Does this closing of the scientific circle threaten important aspects of our self-conception? And how can a scientific worldview justify science itself?

A natural worry about taking the sciences seriously, in exploring scientific practice and using it as a guide for philosophical reflection, is that it will drive out not only the supernatural, but also other things that matter to people (even to people who have abandoned any belief in deities or other transcendent entities). Does science provide a picture of nature as "disenchanted"—as alien and lacking in purpose or value—and can we live with a picture like that?

These are arguably the most serious questions to arise within the philosophy of science, and it is worrying that traditional philosophy of science pays so little attention to them. In the last section, we tried to explain the ideal of methodological naturalism as independent of any wholesale rejection of supernatural entities (although we think there are powerful reasons to suppose no such entities exist). We now want to face up to the possibility that a commitment to the stance of methodological naturalism generates a vision of a world without purpose or value.

During the rise of early modern physics, as we saw earlier (Chapter 1), the idea that inanimate objects exhibit purposeful behavior was strenuously challenged. Galileo was just one among many who treated with withering scorn the Aristotelian idea that motion could be explained in terms of objects seeking their natural places. Effectively, teleological notions (the ascription of purposes and goals) were driven out of the physical sciences. They were retained, however, in accounts of the living world. So, for example, the turning of plants toward the sun was still sometimes seen, even in the eighteenth and nineteenth centuries, in terms of the goal of obtaining light. Indeed, in the discussion of the unity of science earlier in this chapter, we contended that biology, even molecular biology, continues to use functional concepts. Is talk of goals, purposes, and functions scientifically legitimate?

Darwin is sometimes interpreted as eliminating talk of purpose from the biological sciences. That interpretation gains support from one important consequence of his evolutionary picture. Taken as a whole, the history of life displays no overarching purpose: In particular, it is not "headed toward" the emergence of intelligent organisms (among which people tend to count our own species as especially distinguished). Yet there is also a sense in which Darwin allows teleological and functional conceptions—and domesticates them. The turning of the plant toward the sun need not be viewed as some odd effect of a future goal (as if the later state of the flourishing, well-illuminated plant could exert an impact on its current motions). Instead, a Darwinian analysis can suggest that there is some heliotropic mechanism, and that this mechanism has become prevalent in many kinds of plants because ancestral variants with that mechanism (or some primitive version of it) were better able to succeed in the reproductive struggle. Because the mechanism enabled past plants to reach the state of illumination, that mechanism is found in their descendants. This suggests a way of understanding teleology in the naturalistic terms of ordinary causes.

Recent discussion of this possibility has focused on the concept of *function*, for functional explanation is the main form of apparently teleological explanation still in active use in the sciences, and one that seems difficult to do without. The nature of function is particularly important because if biological functions are teleological in any real sense, they hold out the promise of being the most basic form of teleology—the place where teleology comes into the natural world, and the source from which more impressive teleological phenomena such as human planning might be derived.

A functional explanation answers the question why something exists or has the characteristics that it does by adverting to its function. This can be understood as a replacement for traditional teleological explanation. Why do mammals have hearts? The classic teleological response, "To pump blood," is replaced with an ascription of function: "The function of the heart in mammals is to pump blood." If we can then give a straightforward naturalistic account of what it is for a biological item to have a function, naturalists have thought, we might find ourselves in possession of the key to understanding not only the more full-fledged forms of teleology appearing in conscious human purposes, but also related phenomena such as meaning and value.

Many philosophers have been attracted by an *etiological* approach to the concept of function. With respect to living things, we can say that some structure or organ *has the function of doing some particular thing* to express the thought that *past organisms with the structure or organ (or some precursor to it) gained an advantage in the struggle for reproduction because the structure or organ enabled them to do that very thing.* That is one sense that can be given. Yet, in many instances, biologists know rather little about the past environments and the specific ways in which natural selection has operated. They talk about the functions of constituents of cells or of individual molecules by attending to the causal differences those constituents and molecules make. In the background they recognize some general requirements that cells, organs, and organisms must meet if the organisms are to be successful in a Darwinian world. So, for example, molecular biologists might describe a particular enzyme as having the function of repairing DNA, without supposing that they are in any position to say much about the appropriate phase of molecular evolution. They recognize that the molecule has the effect of repairing DNA, and see this as a function because they view the effect as making a positive contribution to the life of the organism, where the notion of positive contribution is understood in terms of making some unspecified contribution to reproductive success. A sober scientific worldview does not deprive us of the possibility of ascribing functions and goals, to the extent that these can be reconstructed in the ways just envisaged.

The possibility of thus making naturalistic sense of functions and goals also opens up new ways of thinking about language and cognition. Some objects (or states of objects) have an odd feature: They somehow point to or are about something else. Sometimes this relationship takes the form of full-blown representation or linguistic meaning; sometimes it is a more minimal kind of indication. Thus the word *cat* (i.e., the ink on this page making up the printed word *cat,* or the sound waves created when somebody says "cat") points to cats; a portrait (even a very bad one) represents the person of whom it is a portrait; and the state of your brain as you think about Shakespeare is related in some special and mysterious way to a particular person long dead. Some nonhuman phenomena are similar: A bird's warning call seems to mean something like "Hawk!"; a bee's dance seems to represent the path to a source of nectar; and it is common to describe stretches of DNA as *genes for* certain traits in an organism. These cases are debatable, but the general phenomenon is not. What is the relationship that allows one object or state to represent or point to another in this way?

A first thought might be that there must be a causal link between them that allows the state of the first object to track or indicate the state of the second. But this is not enough, for many objects are accurate indicators without being representations: Smoke indicates fire, but does not represent fire. On the other hand, as we have noticed, representations can be good or bad, and this means that they sometimes fail to track their targets accurately. A promising solution is to suppose that representations have the function of tracking or indicating the states of their targets—a function that they can fail to perform. Thus meaning, as well as purpose, can be given a naturalistic interpretation.

If we can thus make naturalistic sense of functions, purposes, and meanings, can we also validate talk of values? Here, the issues are far more complex, and we shall only be able to indicate a possibility in barest outline. We begin by rejecting one way of trying to ground our values that, although perennially tempting, is notoriously fallacious.

Ever since Darwin, thinkers have been tempted by the thought that what is valuable is just what has emerged under natural selection. So, in the late nineteenth century, social Darwinists mounted ambitious defenses of rapacious capitalism by suggesting that greedy businessmen exhibited traits that had been refined by natural selection. A hundred years later, some would-be Darwinian ethicists have urged that the central human duty is to spread our DNA, and that people should not copulate with their siblings because of the maladaptive consequences of inbreeding. These crude proposals fail for the obvious reason that there is no reason to think of the blind processes of natural selection as generating traits that are valuable. Long before Darwin's day, Hume had pointed to the basis of the trouble when he inquired how his contemporaries moved from descriptions of facts about the world to prescriptions for human conduct, and the point was reiterated by Darwin's great champion, T. H. Huxley, when he noted that the fact that a characteristic has emerged from our selective history leaves us with the choice of whether to cultivate it or try to eradicate it.

If values are to be reconciled with a scientific worldview, the reconciliation must come, we believe, from recognizing that *valuing* is something people do. We conjecture that the ability to restrain our conduct and deliberately to orient ourselves toward particular goals was an important step in human evolution, one that enabled us to live under conditions of greater cooperation and harmony than those enjoyed by our closest evolutionary relatives. The societies of chimpanzees and bonobos are fragile, constantly afflicted with tensions that must be resolved through time-consuming activities of mutual reassurance. Our capacity for language enabled us instead to talk with one another, to reach agreement on rules of behavior and on goals for ourselves, both individually and collectively. Out of a long period of experimentation come the practices of valuation that we can discern in the past 5,000 years (since the invention of writing). We *make* values, and we have been doing so for a very long time. The practice of valuation was born in societies that were probably as fragile as those of the chimpanzees and bonobos, and it discharged the function of improving our capacities for acting peaceably and altruistically toward one another. That practice has evolved in complicated ways, and it remains unfinished. It is, if you like, a human invention, but it is so deeply embedded in our lives that it is not something we could do without.

This is, as we have suggested, only the barest outline of an approach to a difficult issue. We hope that even so brief a proposal will at least support the view that there is a genuine question here, and that a scientific worldview is not inevitably compelled to discard something central to human life.

Naturalizing Knowledge

We turn finally to an apparently paradoxical aspect of naturalism—what happens when naturalists turn the lens of science on the making of scientific knowledge itself. If naturalists seek to understand or justify science by means of science itself, aren't they reasoning in a circle? And how can an epistemology grounded in mere description do the normative work of distinguishing better from worse ways of choosing what to believe?

Philosophers have often taken a strong line on the second of these questions, arguing for a sharp distinction between psychology and logic. Psychology investigates the way humans actually reason, in all its messy imperfection. Logic treats the abstract relationships among propositions that make reasoning possible, and the principles of good reasoning that follow from them—it reveals not how we do reason but how we should. Because the theory of knowledge has the normative aim of helping us distinguish better from worse ways of choosing or supporting our beliefs, philosophers have sometimes concluded that logic is of central importance to epistemology, whereas psychology is irrelevant.

To understand how naturalism about knowledge can be a tenable philosophical position, it is important to distinguish two different philosophical projects. Some philosophers (prominent in the history of the subject) have wanted to start by making no empirical assumptions, and to arrive at methodological principles that could be used to determine whether a particular belief should count as a piece of knowledge. They have thought that, independently of the way our world is actually constituted, there are guidelines to be followed in the search for truth. Logic tells us, for example, that we are going to be embroiled in falsehood if we believe contradictory things. That is good to know, but it does not take us very far. Can we achieve similar useful principles that might go further in guiding our empirical conclusions?

Naturalists think that cannot be done—at least, not by sitting in
chair and thinking very hard. So they change the character of the task. In
of seeking a set of perfect methods from scratch, they recognize that inves
gators start with a collection of methods and guidelines they have inherited
from their predecessors. Those methods and guidelines support claims that
particular statements express bits of knowledge. Using those items of putative
knowledge, it is possible to reflect on the reliability of some of the methods
used to validate knowledge claims. Current wisdom in physiology or psychology or physics might support the idea that a method traditionally endorsed
is in need of refinement—that as it stands, it is less reliable, less likely to validate true beliefs, than an alternative. Using bits and pieces of science, it is
possible to modify and improve the standards and methods deployed within
science. Instead of generating perfection from scratch, epistemology is recon-
ceived as a *meliorative* project: a venture dedicated to improving what we
currently have.

The new, more modest, aim is to identify principles that can guide us in
improving our epistemic situation—the beliefs we hold and the methods we
use—given our particular epistemic goals, the kind of knowers we are, and the
kind of world we live in. Empirical investigation of many aspects of human
belief can contribute to this sort of meliorative epistemology. Studies of
human cognitive psychology, neuropsychology, and social psychology, of the
history of science and of human evolutionary history, all have a part to play in
helping us to identify processes that reliably lead to true belief for thinkers like
us, and to avoid errors to which we are predictably prone.

But isn't this to engage in circular reasoning, to deploy science in its own
defense? Circularity would indeed bedevil this epistemological approach if
the aim were to justify, from some standpoint without any empirical presuppositions, the methods of science and the knowledge claims they validate. In
fact, however, a more appropriate geometrical image is the spiral: Using some
of what we think we know, we revise some of our methods, we deploy the
amended methods to modify our beliefs, and the amended beliefs then start a
new round of scrutinizing our methods. At each stage, we try to discover
more reliable instruments for the attainment of truth.

Empirical studies of the capacities and processes underlying our beliefs
sometimes suggest that our epistemological situation is worse than we usually
suppose. In recent decades, skeptical worries about the justification of scientific knowledge have garnered concrete backing from several sources—
including evidence from cognitive science and (as we shall see in Chapters 4
and 5) from the history and sociology of science. This evidence suggests that
human knowledge is ineluctably shaped and limited by a variety of constraints
and biases rooted in our evolutionary history, our cognitive apparatus, and
our social nature, and thus that serious challenges confront any epistemology

on that humans can use their experience reliably and
rm true beliefs about the world. Some philosophers
ges as grave threats to the legitimacy of science. But
ituralism can meet, for they arise from naturalistic
een not as threats to the meliorative project in episte-
utions to it. The meliorative project aims to evaluate
emic situation in light of our best knowledge of our-
-knowledge that surely must include whatever we can
learn about the implications of our evolutionary inheritance and human
social psychology for the reliability of our science. The best epistemic options
available to us can indeed be called into doubt, and so too can the sources of
evidence revealing our limitations and biases, but this indiscriminate skepticism is irrelevant to a meliorative epistemology, which seeks not absolute justification but comparative improvement. As we learn more about the world
and our place in it, we learn about better ways of learning.

Suggestions for Further Reading

A classic statement of the idea of the unity of science is provided in Hilary
Putnam and Paul Oppenheim, "The Unity of Science as a Working Hypothesis" (originally published in *Minnesota Studies in the Philosophy of Science*
Volume II, Minneapolis: University of Minnesota Press, 1958). Ernest Nagel's
account of theory reduction is elaborated in Chapter 11 of *The Structure of
Science* (New York: Harcourt Brace, 1961); Hempel provides a simplified
account in Chapter 8 of *Philosophy of Natural Science*. The difficulties of
reducing classical genetics to molecular biology are discussed in the first
chapter of David Hull, *Philosophy of Biological Science* (Englewood Cliffs, NJ:
Prentice-Hall, 1974) and in the first two chapters of Philip Kitcher, *In Mendel's
Mirror* (New York: Oxford University Press, 2003). Nancy Cartwright's objections to unified science are forcefully presented in *The Dappled World*
(Cambridge, UK: Cambridge University Press, 1999); see, in particular, the
first chapter. Disunity is defended from a different perspective in Dupré,
The Disorder of Things.

For David Hume's enduring arguments on induction and causation, see
Hume, *An Enquiry Concerning Human Understanding* (1772)—an excellent
edition is the one by Tom L. Beauchamp (Oxford, UK: Oxford University
Press, 1999). For some of the contemporary approaches to causation, see
Spirtes, Glymour, and Scheines *Causation, Prediction, and Search*;
David Lewis, "Causation," *Journal of Philosophy*, 70, 1973, 556–67; Wesley
Salmon, *Scientific Explanation and the Causal Structure of the World* (Prince-
ton, NJ: Princeton University Press, 1984); Phil Dowe, *Physical Causation*

(New York: Cambridge University Press, 2000); and Christopher Hitchcock "Of Humean Bondage," *British Journal for the Philosophy of Science*, 54, 2003, 1–25. For a thorough critical discussion, see Nancy Cartwright, *Hunting Causes and Using Them* (New York: Cambridge University Press, 2007).

Michael Ruse provides an extremely clear account of the formulation and acceptance of Darwin's ideas in *The Darwinian Revolution* (Chicago: University of Chicago Press, 1979). The evolutionary account of functions was pioneered by Larry Wright, "Functions," *Philosophical Review*, 82, 1973, 139–68. It was further developed by Ruth Millikan, who applied the approach to issues of semantics, in *Language, Thought, and Other Biological Categories* (Cambridge, MA: MIT Press, 1987). The key papers that shaped the debate that followed (and continues) comprise Colin Allen, Marc Bekoff and George Lauder (Eds.), *Nature's Purposes: Analyses of Function and Design in Biology* (Cambridge, MA: MIT Press, 1998). For attempts to elaborate a naturalistic approach to values, see Frans De Waal, *Primates and Philosophers* (Princeton, NJ: Princeton University Press, 2006); Patricia Churchland, *Braintrust* (Princeton, NJ: Princeton University Press, 2010); and Philip Kitcher, *The Ethical Project* (Cambridge, MA: Harvard University Press, 2011).

For discussions of how to ascribe psychological states to nonhuman animals, see De Waal, *Primates and Philosophers*, and Derek C. Penn and Daniel J. Povinelli, "On the Lack of Evidence that Non-Human Animals Possess Anything Remotely Resembling a 'Theory of Mind,'" *Philosophical Transactions of the Royal Society B*, 362, 2007, 731–44. A trenchant discussion of the attempt to resist claims that smoking causes lung cancer is Robert Proctor, *Golden Holocaust* (Berkeley: University of California Press, 2012). Stephen Jay Gould's classic discussion is the title essay of *The Panda's Thumb* (New York: Norton, 1980).

CHAPTER 4

Science, History, and Society

More than Anecdote

Our focus so far has been on the content of the sciences. We have largely ignored the contexts in which scientific work is done. It is, however, a banal fact that the sciences have histories, that those histories intertwine with the development of the societies to which investigators belong, and that today's science is deeply enmeshed in a web of social and political relations. Philosophers have sometimes written as though society and history do not matter, and have often included historical references in their writings merely as decorative anecdotes. In 1962, Thomas Kuhn's *The Structure of Scientific Revolutions* challenged that common philosophical attitude.

Kuhn's work is best understood in juxtaposition with a perspective we dub the *Unkuhn view* of the history of science. According to the Unkuhn view, the recent history of the sciences has shown clear marks of rationality and progress. To be sure, in the more distant past, more or less benighted inquirers fumbled with quaint ideas, but, for each of the mature sciences, there arrived a moment when it came of age—for physics in the seventeenth century, for chemistry and geology in the eighteenth century, and for biology in the nineteenth century. From this point on, cumulative progress was the rule. At the observational level, the sciences piled up well-grounded results about observable phenomena. The process was not quite so smooth at the theoretical level, where corrections sometimes had to be made—even the great Newton had to be refined by the great Einstein—but the correcting theories typically revealed why their predecessors had been worthy approximations. Moreover, the entire process was subject to the governance of reason. Controversies were settled by making observations of nature, abandoning those hypotheses that were falsified (or strongly undermined), and accepting those that were confirmed.

We know of no philosopher who explicitly presented the Unkuhn view as a general account of the history of the sciences. Before the late 1950s (when not only Kuhn, but also Paul Feyerabend, N. R. Hanson, and Stephen Toulmin inaugurated a "historical turn" in the philosophy of science), history did not seem sufficiently important to merit the development of any systematic account. The elements of the Unkuhn view are, however, evident in much contemporary popular writing about science, and discernible in the major works of the analytic project. They pervade the underlying motivation for that project, in which science is taken to be the epitome of reason and progress, worthy of emulation by other forms of inquiry.

Kuhn argued that a more thorough and more extensive look at the history of science would reveal complexities and challenges. His influential monograph proposed that the historical development of the sciences follows a distinctive pattern. In the beginning is *pre-paradigm* chaos, a phase marked by lack of agreement on fundamentals, in which many radically different schools compete. Out of this emerges something entirely different, an activity Kuhn called *normal science*. Scientists working under normal science—as most scientists do all the time, and all scientists do most of the time—operate with a *paradigm* that identifies the puzzles they are to solve, and that sets standards for the solutions they propose: A central task for ancient astronomy, for example, is to represent planetary orbits as combinations of circular motions. Eventually, however, normal science breaks down. A puzzle defeats the concentrated efforts of the most talented scientists, and comes to be seen as something more recalcitrant, as an *anomaly*. This precipitates a new phase, that of *crisis,* in which alternatives to the dominant paradigm are sought. Proposed rivals then confront the older paradigm, and, if one of them attracts the allegiance of most members of the scientific community, there is a *revolution* that institutes a new normal scientific tradition.

Two parts of this account particularly exercised philosophers. First, Kuhn suggested that it was hard to understand the revolutionary debates among competing paradigms as occasions in which rationality triumphed (at least according to the conceptions of rationality favored by philosophers of science). Second, although he acknowledged the power of the notion of scientific progress, he argued that it was difficult to make sense of cumulative progress across scientific revolutions. In these respects, Kuhn's historiography radically undermined the presuppositions of traditional philosophy of science. Some people have never gotten over the shock.

Frameworks and Revolutions

Many scientists reacted more positively, discovering in Kuhn's monograph a sensitive understanding of their practices that had been lacking in the austere logical specifications offered by the analytic project. Instead of focusing on the theories scientists learn, and conceiving of these as sets of statements,

Kuhn talked suggestively of *paradigms*. (He derived the word from discussions of grammar, in which an example is given to illustrate a point: Students learn the conjugation of Latin verbs, for example, by reciting "*amo, amas, amat.*" He later regretted his choice of term, but it was too late; in a range of usages that answer rather variously to his intentions, in all sorts of contexts people talk about "paradigms" and "paradigm shifts.") The point was that scientific training is a kind of apprenticeship, and that the apprentices learn far more than a list of statements. Among the things they acquire from their training are a set of skills, both practical and more purely cognitive. Scientists know how to set up and work various pieces of apparatus, how to evaluate the results of experiments; they know which are the questions that research in their field should address, and the likely lines along which answers are to be sought; they know which of their colleagues to consult about which issues and where to order appropriate materials. A paradigm might thus be thought of as a *framework*, something within which daily scientific practice is carried out, and that includes, among other things, judgments about what is worth doing and what is valuable. (Scientists routinely assess future projects for themselves, and the investigations of their fellows, in these terms.) Kuhn's choice of term was meant to indicate that this framework can sometimes—perhaps even typically—be conveyed by pointing to a particular piece of scientific research that is exemplary for the field. You are to emulate the work of Isaac Newton or Christiane Nüsslein-Volhard, just as the fledgling Latin scholar imitates the grammatical paradigms she has heard.

To practice *normal science* is to take up one of the questions your field takes as significant—one of its existing puzzles—and attempt to solve it. An evolutionary geneticist might tackle issues about the natural selection of traits that seem to be severe handicaps (e.g., the peacock's tail); a geologist might seek an account of the formation of a mountain range in terms of the motions of tectonic plates. If no satisfactory answer emerges from these scientists' efforts, this is not taken to falsify the central theoretical principles that were put to work in their attempted solutions. Failure reflects not on the paradigm, but on the individual scientist.

The account of *scientific revolutions* is best understood in light of Kuhn's ideas about normal science. A crisis occurs when a particular puzzle defeats the best scientists, time after time: At some point, colleagues stop faulting one another for lack of ingenuity, and start worrying about their shared framework. (One of Kuhn's major examples is the persistent difficulty faced by medieval astronomers as they struggled with the orbits of the planets; if a paradigm appears successful in other respects, it may take a very long series of failed attempts to solve a particular puzzle before it is reasonable for people to seek an alternative paradigm.) At this point, some scientists are inspired to conceive alternatives, where these are not simply sets of statements but rival frameworks. Initially, of course, any such new framework will have nothing like the track record of addressing puzzles enjoyed by the older scientific tradition. If it is to be adopted, that must be the result of an assessment of its promise. Kuhn sometimes puts the point provocatively by describing the decisions as based on "faith."

Characterizations of this sort (which, predictably, raised philosophical hackles) rest on several considerations. On Kuhn's account rival paradigms are "incommensurable." This analogy from mathematics (rooted originally in the Pythagorean discovery that $\sqrt{2}$ cannot be expressed as a fraction) is developed in three ways. First, the languages of different paradigms are not straightforwardly translatable into one another: Allegedly, you cannot express the Copernican concept of planet in the language of Ptolemaic astronomy, nor Lavoisier's concept of oxygen in the language of earlier chemistry. Second, proponents of rival frameworks will "observe the same phenomena differently": Standing in the same places at the same times, they will offer different observational reports (Tycho Brahe and Kepler stand on a hill watching the dawn; Tycho sees the sun rise; Kepler sees the Earth rotating so that the sun comes into view). Third, frameworks differ in their judgments of value, specifically those about what problems are worth solving and what the standards for successful solution are. Prior to the seventeenth century, investigators sought teleological explanations for physical processes (thinking, e.g., of motion as directed toward "natural places"); supporters of the various new systems of mechanics abandoned those questions. These three forms of incommensurability entail that an aspiring new paradigm must suffer not only from a relatively scant track record, but also because scientists find it difficult to understand its hypotheses, endorse what its champions claim to observe, and accept its views about what matters in the field. Small wonder, then, that scientific revolutions take a long time. Kuhn's account has the undoubted historical merit of freeing us from the misconception that the opponents of views that eventually triumphed were blind dogmatists, unable to swallow what should have been patent to any unbiased observer.

In response to this challenging picture of the history of the sciences, philosophers have offered attempts to enrich our understanding of progress and rationality. It is possible that Kuhn was correct to demolish the Unkuhn view of history, but that the import of his attack is that we need a more sophisticated treatment than the one delivered by older ideas about theories and their confirmation. That possibility can be explored by examining the three types of incommensurability he claims to discern.

Conceptual incommensurability arises from the fact that many scientific classifications have presuppositions at odds with those made by rival approaches. The pre-Copernican concept of a planet arose from the observation of heavenly bodies that exhibit wandering motions when observed from an

Earth assumed to be stationary; Copernicus, by contrast, thinks of planets as bodies that orbit the sun. Lavoisier proposes that oxygen is an element, absorbed in gaseous form when things burn; for the chemical theorists with whom he disputed, combustion is a process of emission, not absorption— when things burn they give off phlogiston. Does this mean that participants in revolutionary debates are inevitably talking past one another? Communication may indeed be difficult, because simple ways of representing the views of others go awry, but it is not impossible. Sixteenth- and seventeenth-century astronomers overcame the mismatch between concepts of planet by reaching agreement on the individual heavenly bodies they intended to assign to this category. Joseph Priestley, one of Lavoisier's principal opponents, was able to understand that the gas his rival (and friend) called "oxygen" was the substance he referred to as "dephlogisticated air"—and the two men communicated well enough to offer one another descriptions of potential experiments (see "Priestley and Lavoisier Discuss Combustion"). Although Kuhn was right to uncover important conceptual changes across frameworks, the differences do not inevitably produce misunderstandings that would stultify debate.

✍ Priestley and Lavoisier Discuss Combustion

During the middle of the eighteenth century, the most popular approach to chemical phenomena was grounded in an account of what happens when substances burn. Champions of this approach began from the thought that things that can be burned share some common ingredient, and they called the shared substance *phlogiston*. So, the story went, combustion is a process in which phlogiston is released to the surrounding atmosphere.

Beginning in the 1770s, the French chemist Antoine-Laurent Lavoisier conducted a series of experiments (in which he was assisted by his wife—her exact contribution is unknown). Early on, he was able to show that combustion leads to an increase in weight, and he concluded from this that, instead of the release of phlogiston, something is absorbed from the atmosphere. Lavoisier named the substance—the gas—absorbed *oxygen*.

Lavoisier was in communication with a group of British chemists who shared the orthodox phlogistonian perspective. The principal member of that group was Joseph Priestley. Although a very few phlogistonian theorists responded to Lavoisier's discovery of the weight gain by hypothesizing that phlogiston has negative weight, Priestley and his friends adopted the far more common (and plausible) view that combustion involves both the

(cont).

release of phlogiston and the absorption of something else. The rival accounts of combustion can be represented as follows:

Lavoisier: X (*heated*) + air → oxide of X + (air minus oxygen)
Priestley: X (*heated*) + air → calx of X (= X minus phlogiston + Y) + (air minus Y plus phlogiston)

Priestley discovered that when he gently heated the red calx of mercury, he obtained a new gas, one that would enable small animals to breathe and that would support combustion. He envisaged the reaction as follows:

Red calx of mercury + (*gentle heat*) → mercury + dephlogisticated air (air minus phlogiston)

He reported the experiment to Lavoisier, who then performed it for himself. On Lavoisier's view:

Red oxide of mercury + (*gentle heat*) → mercury + oxygen

Kuhn correctly sees that translation from Priestley's language into Lavoisier's, and vice versa, is not easy. From Lavoisier's perspective, there is no such substance as phlogiston. Yet he understands Priestley's term *dephlogisticated air*—and on the basis of that understanding he can do the experiment and isolate that air (which he calls *oxygen*). In communicating with Priestley, he cannot think of dephlogisticated air as meaning "the gas you get when you take phlogiston out of the air"—for he does not think there is any phlogiston, and hence does not think that there is something you get when you remove phlogiston from the air. He has to translate Priestley's reports in a very selective way. Sometimes, when he reads or hears *phlogiston,* he concludes that Priestley isn't talking about anything real at all. But sometimes, when Priestley talks about dephlogisticated air, for instance, he understands that his colleague (and friend) is talking about oxygen.

You probably do not believe in the Tooth Fairy, so if you had a friend who made lots of reports about the Tooth Fairy, you would treat them as not being about anything real. Now imagine that your friend starts talking about the Tooth Fairy's mother-in-law. From various clues, you might figure out that this friend had a very particular person in mind. After that, you would no longer suppose that the meaning of *the Tooth Fairy's mother-in-law* is fixed through the obvious descriptive phrase, but that your friend is using the term to talk about a particular real person. You would be proceeding as Lavoisier did with respect to Priestley.

(cont).

In both cases, the languages are at cross-purposes, and that makes translation sensitive to context. Kuhn's conceptual incommensurability recognizes the difficulties of smooth and uniform translation. It doesn't show, however, that "communication across the revolutionary divide" is impossible, or partial, or even particularly difficult. In fact, the British phlogistonians and the (growing) group of chemists who adopted Lavoisier's views understood one another very well.

Kuhn was also correct in recognizing that proponents of different paradigms will naturally report their observations in language that carries substantive theoretical presuppositions. Yet it does not follow that they are doomed to disagree about the evidence of their senses. Tycho may be inclined to describe the dawn by claiming that the sun is beginning its diurnal round, whereas Kepler is equally drawn to the suggestion that the Earth's horizon is sinking so as to bring the sun into view. Both can agree, however, that the angle separating the solar disk from the horizon is increasing (see "Tycho and Kepler Observe the Dawn"). There is an important general epistemological point here. There is no pure observational vocabulary, no set of terms that will record exactly the content given in perceptual experience. All observation is "laden with theory," since to observe is to bring our experience under concepts, and all concepts have presuppositions. Nevertheless, despite the fact that there is no pure observation language to which scientists can retreat to resolve their differences—no bedrock of uncontaminated observation on which they can stand—proponents of different frameworks can conceptualize their experiences in ways that involve only presuppositions they hold in common. (To say that there is no substance that cures all diseases—or all disputes—is consistent with supposing that for each disease, or dispute, there exists a cure.)

ᥝ Tycho and Kepler Observe the Dawn

Tycho Brahe was a late sixteenth-century astronomer whose observations of the planets were particularly precise and accurate (by the standards of his time). He also proposed a compromise between the traditional Ptolemaic system and the Copernican view: He supposed that all the other planets revolve around the sun, but that the sun itself revolves around the Earth. Johann Kepler served for a while as his assistant, and respected Tycho's

(cont).

observations so much that he demanded that his own theoretical account of planetary motion conform to them.

Imagine the two men watching the sun rise. If their reports of what they saw embodied their different theoretical perspectives, the conversation between them would go like this:

Tycho: Look! The sun is beginning its daily motion around the Earth.
Kepler: I see. Actually, the Earth is rotating, and bringing the sun into view (as it does each morning).

These *observation reports* are explicitly "laden" with theory. But, of course, neither of them has to talk in this way. They can—and probably would—report what they see differently.

Tycho: Look! We can now see a sliver of the sun.
Kepler: Yes. More of the solar disk is becoming visible.

These descriptions are less laden with theory, and, in particular, they have shed the particular theoretical ideas on which the two men differ. But they still presuppose some claims about nature. Tycho talks of the sun as an enduring object, and Kepler adds the idea that it is round. Could they achieve reports that were entirely free of any theoretical commitment?

You might think so. Perhaps the exchange would begin like this.

Tycho: I'm now seeing a slightly rounded yellow patch above the elongated green patch, and the area of the yellow patch is increasing.
Kepler: Me, too.

At this point, any supposition that the sun is an enduring object (round and distant) has been dropped. Notice, however, that the observers continue to take for granted the legitimacy and applicability of various concepts of shape and color. These, too, have presuppositions that might be called into question. So, although this highly stylized language involves less theory, it is still theory-laden.

Kuhn accepts a general philosophical point: Because all observation reports must use concepts, all such reports are theory-laden—there is no "pure observation language". It does not follow, however, that different observers cannot share observation reports. Observations can be recorded in language that only presupposes theories the observers agree on. The second dialog abstracts from the theoretical differences between Tycho and Kepler, and formulates what they see in language that presupposes only what they hold in common—the "theory" that the sun is an enduring, round, distant object.

Kuhn's first two notions of incommensurability are important, but do not ultimately doom the hopes of reasonable resolution of revolutionary debates. The third, however, poses a deeper challenge. As we have seen, paradigms (or frameworks) involve commitments to value judgments. Scientists working within rival paradigms may disagree about the problems that it is important to solve, and diverge on the standards for solution. Moreover, as Kuhn emphasizes, paradigms are never completely articulated, never resolve all the puzzles they set. In a revolutionary debate, champions of the newer paradigm are likely to point to the sequence of strenuous efforts, undertaken by well-qualified practitioners, to tackle what initially appeared to be just another puzzle and is now recognized as an anomaly. The traditionalists will point out that the new paradigm lacks the impressive track record of their preferred framework, and that some previously acknowledged questions have been abandoned. How can differences of this sort be settled?

Initially, there is often plenty of scope for divergent choices. When the large changes in the history of the sciences are scrutinized, it becomes easy to understand how people with different temperaments or commitments might reasonably take opposite sides. The idea of *instant rationality* on these occasions is a philosophical myth. Significantly, revolutionary debate extends over a period of time (about a century in the Copernican example), and during this interval protagonists on each side endeavor to solve some of the puzzles that arise for their favored framework, while attempting to exacerbate the difficulty of those confronting its rival. The dispute about Lavoisier's new chemistry, for example, was worked out over nearly two decades, as chemists of different persuasions (and, indeed, of shifting alliances) devised numerous experiments and responded to them with alternative accounts of the substances involved. As Lavoisier was able to demonstrate that his proposals about chemical composition could accommodate an increasing set of experimental findings, and as the rival suggestions of his opponents encountered difficulty after difficulty, investigators rallied to his program. There was no single experiment or piece of reasoning that was decisive for all of them, but, by the end of the process, almost all of the community of chemists had adopted Lavoisier's framework.

Given that value judgments are involved in this process—and that different frameworks incorporate different sets of values—how can it possibly work? To answer this question, it helps to consider the interplay between value judgments and factual findings in everyday life. Imagine a couple who set out to buy a car. With their resources, the secondhand market seems like the best bet. Initially, one is attracted to a more exciting model (an old Porsche, say), whereas the other favors a more sedate sedan (a younger Toyota Camry). Different attributes are important to them. Yet, as they learn more about their choices, certain things become plain: The Porsche makes strange noises, there are worrying vibrations, signs of rust show around the bumpers, and the odometer revolves in suspicious ways. By contrast, the repair records of the Camry have been painstakingly kept, the ride is smooth, and the body is clean. As these things are learned, the would-be Porsche owner is forced to find a scheme of values that will give precedence to its characteristics over those of the Camry. Smoothness of ride must not count, nor must likelihood of a low frequency of repairs. In the end, perhaps, all that can be done is to point to the attractiveness of the shape—and at that stage the defense collapses.

The analogy suggests that decisions about rival frameworks involve judgment, and we are used to the idea that judgment can be good or bad, thoughtful and informed or casual and ignorant. If revolutionary debates are indeed settled by scientific judgment, that should not be viewed as the abandonment of reason in science. Philosophers have, perhaps, been held too long by the demand that there must be some analog of formal logic that would underwrite all instances of good scientific decision. Part of the importance of Kuhn's work may lie in its freeing us from that constraint, in inviting us to conceive judgment as part of rationality—one not necessarily suited to formalization in the ways for which philosophy has yearned; we regard it as underscoring the conclusions we drew about theories of confirmation in Chapter 2. By elaborating the lines of argument that were offered in settling actual revolutionary debates, historically informed philosophers can expose them as clearly reasonable, without being able to articulate anything like a logic of confirmation or a Bayesian analysis that will characterize them (see "The Devonian Compromise").

✍ The Devonian Compromise

Kuhn's thesis that large debates in science typically involve rival perspectives with partial successes and with unresolved problems is amply borne out by his own historical examples, and many others besides. In the chemical revolution, phlogistonians (like Priestley) claimed that their approach had already answered important questions, whereas Lavoisier (and his allies) emphasized the achievements of the "new chemistry." Their debate involved clashing judgments about which problems are most important to address (this is Kuhn's notion of methodological incommensurability).

Methodological incommensurability also pervades many scientific controversies that occur on a smaller scale than Kuhn's revolutions. One well-studied example is a debate within geology that began in 1834, with the discovery of some anomalous fossils in strata in North Devon (in the southwest

(cont).

of England). Henry De La Beche claimed that the fossils were found deep in supposedly ancient deposits—known as the *Greywacke*—and that they resembled plants already known from Carboniferous strata. Because the Carboniferous was taken to be considerably later than the Greywacke, he concluded that using characteristic fossils to correlate and date strata is a misguided idea.

Roderick Murchison, a staunch defender of the use of characteristic fossils for dating and correlation, contended that De La Beche had made a mistake about the placement of the fossils: They were really at the top of the Greywacke, and the North Devon strata lying above them were actually Carboniferous (so it was no surprise that the "anomalous fossils" were similar to Carboniferous plants). Murchison's proposal faced an obvious problem, however. The largest British Carboniferous deposits were smoothly underlain by a distinctive formation, the Old Red Sandstone, which was absent from the Devon Greywacke. Murchison therefore hypothesized that there must be some discontinuity in the Devon strata, a place where the Old Red was lacking and there was a sudden jump in the times at which adjacent strata had been deposited.

De La Beche quickly modified his views to concede part of Murchison's criticism, allowing that the fossils were indeed at the top of the Greywacke, but he remained firm about the absence of any discontinuity in the Devon strata. The debate persisted for several years, pitting two well-developed positions against one another: Either the topmost Devon strata were relatively young (Carboniferous) and there was a discontinuity lower down, or the Greywacke was a continuous ancient sequence, laid down before the Old Red. The task for both perspectives was to relate the deposits in Devon to other strata, not only in Britain but in Europe (and eventually elsewhere), in some way that was consistent with the observed phenomena. Although Murchison was able to establish some suggestive correlations, his intensive searches for some place at which there was a clear discontinuity in the strata repeatedly failed. For De La Beche and his allies, this was a sticking point—no hypothetical "unconformity" could be accepted unless a discontinuity was found. Murchison, by contrast, emphasized his ability to explain the relations among strata found in an increasing number of places.

Throughout this controversy—the *Great Devonian Controversy* as one of the participants dubbed it—geologists with different opinions disagreed about which problems were most significant. The resolution came with a compromise that allowed Murchison's correlational successes to stand, without the need for a discontinuity in the Devon strata. The age of the Greywacke was fixed between that of the oldest known British deposits (the Cambrian and Silurian) and the Carboniferous. Instead of being as ancient as had been thought, it was coeval with the (apparently very different) Old Red Sandstone. A new geological period had been discovered—the Devonian.

(cont).

This historical case (and others similar to it) has interesting implications for Kuhn's claims. First, it shows how methodological incommensurability can arise in scientific debates less sweeping than those Kuhn studied. Second, it reveals how clash in judgments about what issues are important to resolve need not put an end to reasonable discussion. Finally, it subverts the overly simple conception that paradigms are monolithic, incapable of development and modification, and that, consequently, scientific controversies end with a victory for one paradigm and a defeat for its rival. The Devonian was a compromise, and the final position was one that went unrecognized throughout most of the debate.

The Bogey of Relativism

Kuhn not only raised questions about the rational resolution of revolutionary disputes, but also challenged common assumptions about scientific progress. Viewing the history of a science as a sequence of incommensurable frameworks challenges the idea that we obtain increasingly extensive and accurate pictures of a single world. Instead of showing us as getting ever closer to the truth, history reveals a series of perspectives. Can any of them be seen as superior to others? Aristotle, Newton, and Einstein outlined three large rival visions of the cosmos: If Einstein made progress over Newton, and Newton over Aristotle, in what exactly does the progress consist?

At the end of *The Structure of Scientific Revolutions,* Kuhn posed this question, without providing an answer that satisfied him. Many of those who have been influenced by his work have taken the question to be unanswerable, concluding that the idea of scientific progress is an illusion. Needless to say, the denial of scientific progress has aroused a fierce response, and many philosophers and scientists have vehemently denounced the absurdity of "relativism." Because the debate has been so heated, it is worth looking carefully at how it was generated.

Kuhn was worried about the concept of scientific progress, but far from ready to abandon the idea. His concerns rested on a line of reasoning that has occurred to many people throughout history. At different times and different places, people have formed very different global conceptions of the world: There are contemporary societies that take particular sites to be homes to powerful spirits; in the Aristotelian cosmos there were natural motions of bodies that resulted from their composition out of elements; for some religious traditions the design of nature shows the signature of the deity. When these conceptions are studied carefully, the people who hold on to them turn out to be no less thoughtful or intelligent than those who accept the latest words from the contemporary sciences. From the outside, we might want to say that they are imprisoned by the broad perspective they adopt: In the Aristotelian tradition, astronomers continue, generation after generation, to work on the problem of constructing the planetary orbits out of combinations of circles, even though they bring to their investigations the same qualities of imagination and rigorous reasoning as their Newtonian or Einsteinian successors. Yet it is easy to understand how a similar predicament might affect us, too. Future generations, looking back at our concepts, beliefs, and practices, will very probably consider us to be misguided in important respects, held captive by some large picture that frames our investigations. If our beliefs are true, in the familiar sense of corresponding to the way the world is, there is no way to check that correspondence by stepping outside our own framework to compare the world and our perspective on it. There is no "out of paradigm" experience that could show the world as it is and reassure us that we—unlike the devotees of rival frameworks—are uniquely right.

History, when pursued seriously, raises the bogey of relativism (although some scholars have contended that relativism is not something to be feared, but to be embraced). So, too, does careful ethnography. Beliefs about the world vary widely across places, as well as across times. Anthropologists who spend years articulating the ideas of an alien culture often provide accounts that reveal the reasonableness of responses that once appeared bizarre to the point of absurdity. Understanding the details of variation in belief, across contemporaneous cultures or across historical time, suggests an unnerving symmetry, inspiring worries (or delight?) that "everything is relative." What grounds can we have for saying that our own beliefs, even those we take to be best supported, are probably good approximations to the truth, whereas the beliefs held as best supported by people who differ profoundly from us merely appear to them to be true? What justifies the conclusion that they, and not we, are imprisoned by the broad perspective of a culture?

Relativistic concerns of this sort emerge from attempts to think in a careful and scientific way about human cultures. In the late nineteenth century Franz Boas, often regarded as the founder of modern anthropology, noted that the study of humans by humans poses special problems for empirical science. Anthropologists' own culture can be a powerful source of bias in their attempts to understand other cultures, shaping and constraining their interpretations and even their observations. To minimize the effect of these ethnocentric biases, Boas suggested, anthropologists should strive to understand other cultures "from the inside"; this would provide critical distance on their own culture and help them to guard against parochial biases. Anthropologists should seek to understand and evaluate the beliefs, values, and practices of each culture using only the concepts and standards belonging to that culture itself.

Kuhn's reflections on the history of the sciences were inspired by a similar conception of appropriate method. Even historians who favor different accounts of the historical episodes he discussed (and who reject Kuhnian terminology) are likely to agree on this point: Very little serious history can be done by dividing the scientists of the past into the bold and insightful innovators we now see as having been on the right track, and the dogmatic twits who opposed them. Anthropology after Boas and the history of science after Kuhn share a thesis of *natural reasonableness*: In different eras and in different places, people are equipped with the same cognitive faculties, and those faculties are put to work in similar ways. Natural reasonableness grounds a methodological principle for good investigation. Different beliefs should be explained symmetrically, traced to the operation of the same types of causes. A decade after Kuhn wrote, that principle was made explicit in the work of the Strong Programme in the Sociology of Knowledge (henceforth sociology of scientific knowledge, or SSK): Whether or not beliefs are taken to be true or false, they should be explained in the same terms.

This methodological principle is extremely plausible, although, as we shall see shortly, it needs to be clarified in some respects. The principal question is this: What follows from it? Does the symmetry principle imply that it is impossible to justify claims about whether one belief is closer to the truth than another? Or that the only defensible talk of truth must be relativized to a culture or to a perspective—that what is true for that culture (or for that perspective) is just whatever the members of the culture (or adopters of the perspective) accept? Champions of SSK, and those who have built on SSK to develop other—often more radical—views about the history and sociology of science, have frequently been swept into affirmative answers. They are committed to forms of relativism that alarm many self-styled defenders of science.

The line of thought is apparently straightforward. If you think about rival systems of belief in their own terms, you will appreciate them as reasonable ways of responding to the world, as having the same force for those who hold them as your own system does for you. So you will have no basis for "grading" them, for holding that one is closer to the truth than any other. Because talk of truth without any way of identifying which systems are closer to or farther from the truth would be idle, the concept of truth should either be dropped or identified with what is culturally accepted. The idea of progress is part of a mythology that regards the sciences as uniquely important forms of knowledge. Instead, we should abandon this sort of cultural imperialism and recognize that there are many "knowledges" or "ways of knowing."

On closer inspection, however, the argument is more intricate than it initially seems. Let's start with the symmetry principle itself.

Across the world and throughout history, human beings have arrived at their opinions by using their senses, drawing on their memories, exchanging information with one another, and developing chains of reasoning. We sometimes recognize particular people as specially gifted—they can tell at a taste what ingredients have gone into making a complicated dish, they are able to hear a musical piece and write out the score (Mozart famously did this for a composition kept secret by the papal choir), or they can remember entire texts word-for-word after reading them. When there is a basis for attributing a conspicuous talent of this sort, there is no obstacle to invoking it in explaining why someone believes what she does. Most of the time, however, our accounts of others' views start from thinking of those other people as having the normal stock of human capacities. Crucially, unless there is contrary evidence, we do not take them to wander about the world with their eyes closed or to be seduced by blatantly illogical reasoning (although to the extent that psychologists show how we often make inferential errors we anticipate that those too will afflict the people whose convictions are under study). So there is a plausible core to the symmetry principle: At a very general level of description, all beliefs are formed in the same ways, through sensation, conversation, memory, and reasoning.

Recognizing that fact, however, does not preclude the possibility of finding differences among performances in belief formation. Indeed, differences are identified all the time, especially in educational contexts. When schoolchildren take examinations, they sometimes offer different answers to the same questions. Some of them misremember, or misapply rules they have been taught, or fail to see the right way to approach the problem they have been set. Differences of this sort do not threaten the core of the symmetry idea. Nevertheless, these differences reveal that the fact that all the children go through the same generic processes does not entail very ambitious claims about symmetry among them: It doesn't imply that no child remembers better than any other, or that the children are all on a par when it comes to applying rules, or that they have all had the same educational opportunities. So it looks as though the idea of natural reasonableness only supports a very weak symmetry principle, one that allows for making distinctions among different styles of belief-formation, and thus among systems of belief.

Is this an appropriate analogy? Defenders of the symmetry principle might reply that the analogy overlooks the systematic character of the differences separating large frameworks. If we can distinguish among students' reasoning processes, segregating those who reasoned well from those who made errors of one sort or another, it is because we already agree about the correct answers to the questions posed to them. Scholars impressed by the coherence of the belief systems of different cultures, or of different perspectives adopted during the history of science, might protest that no similar background of agreement is to be found in these cases: Aristotelians were pursuing different goals from the Newtonians and Einsteinians who succeeded them. The more penetrating histories of the sciences offered by Kuhn and his successors show that the appearance that scientists in different contexts face a "common examination" is merely superficial—the major traditions in physics, chemistry, and biology might look as though they are addressing the same questions, but the examination papers are really written in different languages and the standards for correct answers are thoroughly diverse.

The argument for relativism turns out to lean heavily on the theses about incommensurability discussed in the last section. Hence it is worth recalling a principal point from that discussion. Revolutionary debates do often involve different ideas about what questions need to be answered and what scientific projects are significant. But they are won, we suggested, when one of the parties is able to achieve a strong track record of successes with respect to problems it identifies as important, while demonstrating that systematic difficulties confront its rivals in living up to their own standards. If those rivals are to continue to resist, they have to amend the conception of what needs to be accomplished until it collapses into absurdity: To this end, we developed the story about buying a used car, in which the Porsche lover was eventually driven to propose that attractive shape outweighs considerations of reliability and smoothness of ride. That tale has a moral for the discussion here. Merely pointing out that different cultures have different projects, pursue different goals, and hence set different standards does not show that they cannot be compared. The successful pursuit of some projects may turn out to be impossible (you cannot have reliable transportation and the attractive shape), and insisting on some particular goal ("The car must be cool!") may make a mockery of the enterprise of buying a car.

Would-be relativists will not be convinced. They will press their fundamental thought more deeply: Judging a particular scheme of values to be absurd itself reflects parochial bias. However long a list of successes Lavoisier racked up, his phlogistonian opponents could always point to some unsolved problem as being of overriding importance; the Porsche lover's ultimate standard of the beautiful shape only makes a "mockery of the enterprise" if you think that the project of car shopping has something to do with being able to move from one place to another. Relativism only falters if ultimate values that differ from our own are labeled *absurd*. It is exactly that practice of labeling that relativism has diagnosed as parochial, from the beginning.

At this point, champions of science can easily throw up their hands in frustration, impatience, or despair. They shouldn't. What makes resistance to Lavoisier's new chemistry problematic is that Lavoisier and his allies can *do* so much: They can predict how experiments will go, and they can manufacture all sorts of compounds. Similarly, while the Porsche gracefully adorns its owner's driveway, the Camry buyer can undertake all sorts of expeditions. So, you might say, understanding, predicting, and making compounds is a constitutive goal of chemistry, and obtaining reliable transportation is a constitutive goal of car shopping. More generally, predicting the course of nature and intervening in it are constitutive goals of science.

People troubled by the bogey of relativism often resort to a gibe: "Show me a relativist at 30,000 feet!" The underlying point should now be apparent. The sciences help us achieve a lot of practical things. They supply the basis for successful predictions and interventions. That is why they are privileged over other systems of belief.

As the next section shows, however, this is not the end of the complications, or of the argument.

Success, Truth, and Progress

Contemporary molecular biology enables scientists and laboratory technicians to do a lot of striking things. Mapping and sequencing the genomes of many organisms has now become almost routine. On the basis of the data obtained, it is possible to modify the genomes of particular organisms in a vast number of ways: Flies with a mosaic of tissues never found in nature can be created for experimental purposes; genetic material can be inserted into bacteria so that the bacteria will manufacture particular substances, including compounds that are valuable in treating diseases. Day after day, all over the world, activities like these go on, delivering organic products that are useful in a wide range of human projects.

How could all this succeed so brilliantly if the underlying basis on which the work rests, the molecular account of cutting and splicing DNA, the maps, and sequence data, were not substantially correct? It is tempting to declare that if the hypotheses of contemporary molecular genetics were thoroughly false, it would be a miracle for biomedical technology to work in the wide-ranging and reliable ways it does. In recent decades, many philosophers have given in to the temptation and accepted the "miracle" argument for scientific realism.

The miracle argument is most convincing when there is a systematic body of predictions and interventions that can be carried out across a broad range of contexts. Molecularly based biotechnology provides one good instance, but others abound in physics, chemistry, and the earth sciences. Versions of the argument gain force when prediction is extraordinarily fine-grained (as with the ability of quantum theory to achieve a degree of accuracy analogous to measuring the distance between New York and Los Angeles with an error less than the width of a human hair) or when the tasks accomplished are apparently extremely difficult (bringing back materials from Mars or making a fly with a prescribed mixture of male and female tissues). Yet, even in the most impressive cases of scientific success, the miracle argument faces criticism.

Start with an obvious point. What counts as success? Relativists will surely insist that what people take to be successes are affected by their individual aspirations and by the values of their culture. People with very different goals will be quite unimpressed by the ability to direct rockets at distant planets, or the manufacture of bizarre flies. Aristotelians were not much concerned with giving a mathematical description of the motions of projectiles, so Galileo's ability to provide such descriptions would not have advanced their goals—even though Galileo was able to offer recommendations to gunners about how to obtain maximum range for their cannonballs.

The success that is of interest to the miracle argument abstracts from particular human goals. It focuses on the processes of predicting and intervening. To be successful is to offer a prediction or make an intervention that allows a goal—any goal—that was previously unattainable, to be reached. When that can be done systematically, accurately, across situations in which the goal has defied previous human efforts, the miracle argument comes into play.

Of course, at different times and in different places, people sometimes achieve the same goals by alternative means. Consider the agricultural enterprise of developing new strains of crops. The use of genetics, whether the classical genetics of Mendel and Morgan, or the molecular genetics of today, is a recent and localized method of trying to improve quality and yield—it is not the only such method. Generations of farmers and breeders have developed skills that help them reach their goals, without knowing anything about the mechanisms of heredity. Sometimes their practices may be based on explicit beliefs about how to proceed; on other occasions, they have skills they cannot articulate. If the miracle argument applies to the genetic approach to plant breeding, leading to the conclusion that its underlying account of the hereditary processes is approximately true, how does this bear on the practices of people who achieve similar successes in quite different ways?

Scientific realists expect to be able to *explain* the successes of the skillful farmers. When those farmers proceed by using explicit lore, it is anticipated that the rules of thumb they employ will be seen as approximations to special cases that genetic theory identifies. When they cannot articulate their skills, it's supposed that genetics would make sense of the cues to which they respond: They focus on selecting this type of plant or that, because the observable features they prefer are reliable indicators of underlying combinations of alleles that promote drought tolerance or high protein content, for example. If this can be done, successful practices that are not based in genetic theory do not have to be viewed as rivals that would undercut the miracle argument—that is, as incompatible systems of belief with comparable ranges of success—but as absorbed within the far more exact and wide-ranging basis for prediction and intervention that genetic theory supplies. If it cannot be done, there is a genuine worry. For then there *are* two incompatible sets of principles, genetic theory and the practical rival, both of which could claim support from the miracle argument: If we can argue from success to approximate truth for genetic theory, we can do the same for the practical rival, and, because they cannot both be true, the form of argument is revealed to be problematic.

Although nobody has investigated (or could investigate!) all the indigenous practices of plant cultivation, past and present, enough is known to quell concern that there will be a threat to the miracle argument's endorsement of genetic theory. Other examples pose more difficult challenges, however.

For generations, the distribution of water to the rice-growing terraces on the island of Bali has been carried out by an elaborate organization involving local deities and the water temples they are supposed to frequent (see "The Water Temples of Bali"). Western scientists believed that they could replace this baroque rigmarole with something more streamlined. After years of effort on the part of hydrologists, ecologists, and computer programmers, these scientists were forced to recognize the effectiveness of the indigenous organization: Virtually all of the recommendations they made did substantially less well. On the face of it, Balinese practice—a practice permeated by religious ideas—seems to have at least one major systematic success. Should we therefore conclude that its theoretical basis, including the claims about the local deities, is true?

ᥑ The Water Temples of Bali

For centuries, farmers on the island of Bali have successfully cultivated rice. Their agriculture has depended on the construction of a complex system of weirs, dams, tunnels, and canals that direct water from the rivers that flow down the sides of volcanoes into the rice paddies of the lowlands. Water temples occur at strategic points in the irrigation system. As long as the indigenous approach to agriculture was dominant, each temple was visited only for occasional ceremonies, attended by a group of farmers.

In the late 1960s, after the Green Revolution, Balinese agriculture changed. The Indonesian government began a program to improve rice yields, using the apparently promising crops developed by scientists. Farmers were directed to use the new varieties instead of their traditional strains, which took longer to mature. A second stage of the program attempted to reform the system of irrigation, in response to the need to adjust water flows so as to avoid the outbreaks of pests that initially plagued the new agriculture. Despite the use of sophisticated methods of ecological modeling, the efforts to improve on the traditional system of irrigation failed. A report

(cont).

from the Asian Development Bank, published in 1988, concluded, "The substitution of the "high technology and bureaucratic" solution in the event proved counterproductive, and was the major factor behind the yield and cropped areas declines experienced between 1982 and 1985." That report described the Balinese rice terraces as a "complex artificial ecosystem," understood by the local farmers.

There is no doubt that the traditional system of water temples, assemblies, and ceremonies was successful in enabling rice cultivation and that the "high technological" substitute was not. Does that support an inference from success to truth, one that entitles you to believe in the deities whom the Balinese farmers honor in their ceremonies? Surely not. The traditional system works so well, not because there are beings who, when suitably celebrated, make the water holy, but because the ceremonies occur in response to the farmers' careful observation of prevailing conditions—the moisture or aridity of the soil, the density of insect populations—and their assemblies lead to joint action, adjustment of water flows, that benefits all rice growers in a particular area.

In effect, there are two different problems:

1. Find a means of coordinating water flows so as to grow healthy crops, given human observation of local conditions and cooperative social interactions.
2. Find a means of coordinating water flows so as to grow healthy crops without depending on cooperative social interactions.

It should not be surprising that the second problem is harder to solve than the first—or that careful everyday perception and harmonious social coordination can play a crucial role in managing the environment.

Or consider the track record of the sciences themselves. Over the course of previous centuries—skeptics about the miracle argument contend—any number of theories about all sorts of aspects of nature have been accepted for a time, theories that could boast substantial success. By our current lights, almost all these theories are wrong. Hence, it is plain that success has nothing to do with truth.

Is this to overpraise the efforts of past theorists? It is hard to perceive very much interventional or predictive success in the cosmology of the middle ages, or the humoral theory of medicine, or phlogistonian chemistry. But there are more challenging cases. Nineteenth-century chemists achieved remarkable analyses and syntheses of a wide variety of compounds without recognizing the internal structure of the atom. Even more strikingly, the wave theory of light, committed from the start to the notion that any wave motion requires a medium of propagation and so relying on the existence of an all-pervading ether, was able to make entirely unanticipated predictions. Most famously, as Chapter 2 noted, the initial presentation by Fresnel was challenged by the French mathematician Poisson, on the grounds that Fresnel's mathematics of wave propagation would predict the existence of a bright spot at the center of the shadow cast by a small disk; the objection prompted Arago to set up the experiment and to discover the spot, a striking triumph for the wave theory.

Before we leap to the skeptic's judgment that examples like this undermine the thought that success betokens truth, there are two important points that should be recognized. First, theories that are, taken as wholes, incorrect, are often not entirely wrong. Fresnel may have believed, falsely, that wave motion required a medium of propagation, and so invoked an ether, but his mathematical account of the ways in which transverse waves propagate nonetheless lives on in contemporary optics. The account can be freed from his incorrect, but comprehensible, assumption that there had to be a "luminiferous ether." The nineteenth-century chemists divided into two groups, those who doubted the reality of atoms and those who thought atoms were indivisible and structureless. Both, we now believe, were wrong, but many of the atomic formulae and structures they posited have been retained. When truth and falsehood are mixed, the success of the applications of the theory will depend on the elements selected from the mixture. Dramatic predictions or interventions signal the use of accurate parts of the theory. The structure of atoms is quite irrelevant to much successful nineteenth-century analysis and synthesis, and the hypothesis of an ether plays no role whatsoever in the prediction of the Poisson bright spot.

This point enables us to make sense of the Balinese water temples. In effect, the Western scientists were attempting to address a problem with an important social dimension. You can either set up a system with rules for water allotment, and try to persuade people to follow those rules, or you can attempt to find a scheme that will effectively force the distribution you want whatever the participants do. The local Balinese solution takes the first form, with the doctrines about temples and deities playing the familiar role of fostering social cooperation: The shared myths bring people together. Once they have assembled, they exchange information, and their fine-tuned knowledge about water flows inspires the design and use of water courses. The Western scientists tried to solve the second problem, and their schemes were obviously inferior because, under varying conditions, the Balinese could adjust water flows and varieties of rice as required (and, because of their socioreligious commitments, were prepared to do so in a collectively beneficial way). Once this is appreciated, it is easy to see that the example neither undermines the inference from success to truth nor provides a basis for thinking that Balinese religion has a special role in generating successful predictions and interventions. Any effective way of bringing people together to share information and to cooperate would work as well.

Furthermore, there are occasions on which different approaches to a domain of phenomena can be successfully combined with one another. As we noted in Chapter 3, a Newtonian model of motion may work well for explaining or predicting what happens when you drop a cannonball from the top of the Empire State Building, but if what you drop is a twenty-dollar bill, a different model is needed: one from fluid dynamics or even economics. Different theories and models, often not reducible to any common basis, capture different parts or aspects of the world.

There is no reason to suppose that the diversity of models here implies any problem about which of them is closer to the truth. Consider the even simpler and very familiar case of the making and use of maps. Many maps represent the Earth's surface, in part or in whole, but no map represents every aspect of the part of the Earth's surface that it represents; all maps leave out a great deal of information. What they include depends on their particular function. Subway maps, road maps, and navigational charts include very different kinds of information to suit the uses to which they are put. But it doesn't make sense to ask which is true, or even which is closer to the truth. Which one is better depends on what you need it for. (On the other hand, the representations of areas on a Mercator-projection map, or of the relative distances between stations on many subway maps are not accurate, or intended to be accurate.)

There is good reason, therefore, to adopt a pluralist perspective at least in the sense of accepting that many different kinds of representations are needed to make sense of the diversity and complexity displayed by the world's phenomena. The kind of pluralism we are proposing here resembles relativism in recognizing that diverse representations of nature may be equally good in different ways. But it is also compatible with realism.

A second point should also be made in response to skepticism. Why do we think that successful practice redounds to the truth of the ideas used in generating the predictions or interventions? Our grounds do not lie in any deep metaphysical view, but in everyday experience. Often, we find ourselves temporarily prevented from perceiving how things are in some local situation that poses threats or challenges to us. Repeatedly, we have to form representations of arrangements to which we have no direct access and to adjust our behavior in light of what we think is going on, and we discover, again and again, that when the problems we are trying to solve are systematic, or are hard, we need accurate representations to succeed. These situations arise in our dealings with other people, in the causes of disease symptoms, in detective investigations of many kinds, in finding our way when we are lost, and perhaps most purely in those games where we have to hypothesize the distribution of the cards or the configuration of the opponent's pieces. In many such instances, there comes a point at which the veil is lifted, when we can check our anticipatory representation of the situation against the way things are. Throughout our lives we accumulate evidence for the idea that success betokens truth, without ever focusing the question in any explicit or systematic way, and from this arises the conviction that we would not succeed so dramatically if the hypotheses that guide us were thoroughly false.

Our defense of the miracle argument depends on distinguishing those parts of our system of beliefs that are genuinely put to work in generating the successes from the parts that are idle. Sometimes the distinction can be made with relative ease by later analysts, but not by the investigators who are developing and defending a theory. Fresnel could not conceive of the possibility of a wave motion in empty space, and concluded that an all-pervading ether was essential to the mathematical account of light transmission he presented. His successors agreed, setting themselves the task of working out the mechanics of the ether—no less a scientist than James Clerk Maxwell thought that the ether was one of the best confirmed entities in natural philosophy. With hindsight, we can see that the ether was unnecessary (and, eventually, a source of problems), but brilliant scientists, unable to locate the distinction between what was actually doing the predictive work and what was being (wrongly) taken for granted, overestimated the contents of the universe.

The moral is that our attitude toward the conclusions of versions of the miracle argument should be cautious. Truth is frequently mixed with error. People do not always disentangle parts of their overall view, recognizing clearly which ones are being deployed in its successes. In consequence, it is easy to suppose that assumptions irrelevant to the prized achievements are also sound. Like our predecessors, we are probably making mistakes somewhere, failing to distinguish those elements in our views that genuinely feature in our successes from those that are getting a free ride. Unless—or until—we have grounds for specific doubt, we can do no better than to accept the whole, with some confidence that large parts of it will endure, while recognizing that other parts will probably need revision. When researchers claim that the entities introduced by their apparently successful theories exist, their position is that of the author of a thoroughly researched book, who checks every sentence and assents to it, but rightly confesses in the preface that there are probably some residual errors.

Modesty is also commended by another insight from the history of the sciences. One of the striking features recognized already by Kuhn and emphasized by some of his successors is that even successful sciences are often framed by assumptions that later advances will revoke in radical reconceptualizations.

The wave theory of light has given way to quantum electrodynamics, Newtonian mechanics (staggeringly successful by pre-twentieth-century standards) was displaced by quantum theory and the theories of relativity, and contemporary molecular genetics transforms the early attempts of Mendel and his rediscoverers to analyze inheritance in terms of the transmission of particulate factors. In all three instances, the concepts eventually used to generate predictions and interventions lie beyond the horizons of those who reap the early successes. Hence, reflection on the conspicuous success of any area of research should be tempered by the thought that future investigators may adopt a perspective very different from our own. Nevertheless, just as Fresnel's mathematics of wave transmission, Newton's laws of motion and of gravitation, and Mendel's "Rules" are all reframed and preserved within the contemporary treatments of light, motion, and inheritance, it is reasonable to expect the ideas that are put to work in successful prediction and intervention to endure, in some form or another, as the sciences advance further.

Progress Without Truth?

Let us take stock. Worries about relativism led us to consider the popular response that areas of mature science are strikingly successful, and thus likely to be true—a response implicit in the frequent charge that relativists are guilty of bad faith when they travel by airplane or take the medicines prescribed by their doctors. As we have seen, judgments about success and inferences from success to truth are both more complicated than they might initially appear. In this concluding section, we want to look briefly at an apparently simpler option for coming to terms with relativism.

Start with a question: Does the emphasis on scientific success violate the presumption of natural reasonableness, which is such an attractive part of the relativist's perspective? No. We can understand failure to reach the correct view of some recondite aspect of nature without disparaging the acumen of those who fail. Specialist expertise is often a sign of an idiosyncratic interest, rather than of any superiority in cognitive ability. People in different cultures don't arrive at the same complicated theories we hold to be true when those theories are generated by questions whose significance for the other cultures is not particularly salient or urgent. (Our account of the molecular basis of heredity is a case in point.) Further, we might point to a difference in the ways in which the efforts of individuals are coordinated. Since the seventeenth century, Western societies have developed increasingly complex social institutions for the collective pursuit of inquiry about nature, and it should not occasion surprise if the products of such sustained (and expensive) investigations should be more accurate than the beliefs of people who live in societies that lack both the institutions and the resources that underlie them.

So, we might say, people in different societies (or people living at different periods of time) can be equally intelligent and imaginative, but focused on different things. If a topic excites only casual attention in another group's investigations, that group is likely to have less well-developed views about it. By the same token, it is appropriate to take seriously the perspectives current in different societies with respect to the questions of most intense concern to them (this thought will occupy us in Chapter 5).

Natural reasonableness might produce a condition of rough equality among groups (at least insofar as differences resulting from the creation of a class of specialist investigators and of coordinated inquiry do not come into play). Different societies respond to the most urgent problems they face and come up with their own successes. Why not rest content with an understanding of the distributions of success, leaving talk of truth out of account?

This ecumenical idea is attractive, and is readily extended to an account of scientific progress that forswears inference from success to truth. Recall the idea of the "dappled world," offered by Cartwright (Chapter 3). On this view, the sciences march forward by finding or creating pockets of order, providing an expanding set of models that enable people to predict and intervene in ways that matter to them. (Perhaps also to answer the explanatory questions that arise for them—but retaining that thought would depend on divorcing the notion of a successful explanation from the attainment of truth, or of approximate truth.) Progress would then consist in achieving increasingly extensive or powerful models with respect to the tasks of predicting and intervening that people take to be important, subject to the proviso that decisions about which tasks were important were informed by good judgment.

This is a pragmatic approach to progress, on which those who resist the relativist idea that the history of the sciences is no more than a series of changes in fashion can attempt to build further. There is no need for the sweat and strain that defenses of "scientific truth" involve. Kuhn worried that his historical studies left no room for the notion of scientific progress understood as increasing approximation to the truth, and one way of answering that worry is to divorce the idea of progress from claims about truth, in the way indicated in the last paragraph. Yet it is also possible to respond directly to the ancient argument Kuhn revived. According to that argument, there is no "view from outside" in which we can compare the world and our representations of it, and be comforted by their agreement. Nonetheless, there is something interestingly similar we can do: We can observe one another.

People navigate their ways around the environment by having thoughts that represent parts of their surroundings. Views about the relationship between representations and the world they aspire to represent are generated from reflections on the representation-coordinated behavior of others. The thought that you live in a world that is independent of you (and of what you think) arises from the perception that other people come and go in a world that is independent of them. You appreciate that you are no different. Onlookers, tracking you, would conceive of your relationship to the surrounding world as you think about the relationship between other people and their environments.

As we have already suggested, part of ordinary experience consists in understanding how accuracy of representation affects successful action. Imagine yourself watching someone explore unfamiliar territory with the aid of a map. She announces her goals, and provides you with a running commentary as to how she is linking the signs on the map to objects in the vicinity. When she has been equipped with an inaccurate map, or when she is incorrect in associating map symbols with the pertinent objects, she rarely reaches her chosen destination. Wide-ranging and systematic success comes only when her map is accurate and she becomes adept at finding the environmental correlates of the symbols.

The map user's representation of the surroundings directs her actions, and success comes when the representations she achieves are connected to the surroundings in the right way. The general cognitive predicament we all face is similar. Symbols, words, the elements of thought, are linked to—refer to— entities that are typically independent of us and of our thinking, and our representations are accurate or correct when the things to which we refer fit together in the ways we represent them as doing. That is the everyday source of the thought that truth consists in "correspondence to reality."

So, although a minimal notion of progress, understood in terms of increasing predictive and inferential success, might answer Kuhn's worry, we think that ordinary considerations provide grounds for thinking that our successes are based on accurate representations of an independent world, and that scientific progress consists in getting closer and closer to a true account of the parts of it that matter most to us.

Suggestions for Further Reading

Kuhn's ideas are presented in *The Structure of Scientific Revolutions* (Chicago: University of Chicago Press, 1962; 2nd edition, with an important Postscript, 1970). Several of the essays in his later collection, *The Essential Tension* (Chicago: University of Chicago Press, 1977), are valuable guides for interpreting the earlier book; particularly important is "Objectivity, Judgment, and Theory Choice."

For ideas about theory and observation akin to those expressed by Kuhn, see N. R. Hanson, *Patterns of Discovery* (Cambridge, UK: Cambridge University Press, 1958), Chapter 1. A more radical view about incommensurability and the impossibility of resolving scientific revolutions is proposed and defended in Paul Feyerabend, *Against Method* (London: Verso, 1975).

Two important early reactions to Kuhn, which elaborate alternative visions of scientific change, are Imre Lakatos, "Falsification and the Methodology of Scientific Research Programmes," in I. Lakatos and A. Musgrave (Eds.), *Criticism and the Growth of Knowledge* (Cambridge, UK: Cambridge University Press, 1970), 91–196; and Larry Laudan, *Progress and its Problems* (Berkeley: University of California Press, 1977). Later attempts to offer a philosophical account of the growth of science, which respond not only to Kuhn but to the tradition in historical studies of science that he initiated, are Michael Friedman, *Dynamics of Reason* (Stanford, CA: CSLI Publications, 2001), and Philip Kitcher, *The Advancement of Science* (New York: Oxford University Press, 1993).

For SSK and later developments in the history and sociology of science, see Barry Barnes, *Scientific Knowledge and Social Theory* (London: Routledge, 1974); David Bloor, *Knowledge and Social Imagery* (London: Routledge, 1974); Harry Collins, *Changing Order* (London: Sage, 1985); Bruno Latour and Stephen Woolgar, *Laboratory Life* (London: Sage, 1979); and Bruno Latour, *Science in Action* (Cambridge, MA: Harvard University Press, 1987). Some of this work has given rise to unsympathetic (even scathing) critique; for example, Paul Gross and Norman Levitt, *Higher Superstition* (Baltimore: Johns Hopkins University Press, 1994) and Noretta Koertge (Ed.), *A House Built on Sand* (New York: Oxford University Press, 1998). A classic collection of papers from the debate over relativism is Martin Hollis and Steven Lukes (Eds.), *Rationality and Relativism* (Oxford, UK: Blackwell, 1982). Clifford Geertz, "Distinguished Lecture: Anti Anti-Relativism," *American Anthropologist*, 86, 1984, 263–78, gives a ringing statement of the anthropological perspective. Another approach—a distinctive form of pragmatism—appears in many of the essays in Richard Rorty, *Objectivity, Relativism and Truth: Philosophical Papers I* (Cambridge, UK: Cambridge University Press, 1991); we especially recommend "Solidarity or Objectivity?."

The miracle argument (sometimes referred to as the "no miracles" argument) was originally formulated by J. J. C. Smart in his *Philosophy and Scientific Realism* (London: Routledge, 1963). The classic critique, based on the pessimistic induction on the history of science, is Larry Laudan, "A Confutation of Convergent Realism," *Philosophy of Science*, 48, 1981, 19–49; reprinted as Chapter 5 of *Science and Values* (Berkeley: University of California Press, 1984). Arthur Fine's *The Shaky Game* (Chicago: University of Chicago Press, 1986) is a major contribution to debates about scientific realism. For more recent discussion, see Stathis Psillos, *Scientific Realism* (London: Routledge, 1999) and P. Kyle Stanford, *Exceeding Our Grasp* (New York: Oxford University Press, 2006).

Good accounts of the chemical revolution and of debates between Lavoisier and defenders of the phlogiston theory are provided by John McEvoy, "Continuity and Discontinuity in the Chemical Revolution," and, with great lucidity and thoroughness, by Frederic L. Holmes, *Lavoisier and the Chemistry of Life* (Madison: University of Wisconsin, 1985). The views of Tycho Brahe and Johann Kepler are clearly presented in Thomas Kuhn, *The Copernican Revolution*. Martin Rudwick, *The Great Devonian Controversy* (Chicago: University of Chicago Press, 1985), is, in our judgment, the most detailed and multifaceted study of any scientific controversy—and also a fascinating read. J. Stephen Lansing, *Priests and Programmers*, (Princeton: Princeton University Press, 1991) is an authoritative study of the Balinese system of irrigation and of the attempt to improve on it.

CHAPTER 5

Critical Voices

A Mixture of Challenges

Science has always had its critics, and as scientific knowledge and its applications have been extended in ways that bring increasingly far-reaching changes to our world and our understanding of it, critical voices have become louder and more urgent. The challenges that have emerged in recent decades are diverse in origin, and often directed at different targets. Although there is a tendency to hear critical voices as "anti-science," that is frequently to mistake the intentions behind their challenges. Some commentators on the sciences are not concerned with finding flaws in scientific practice, but simply in correcting a prevalent image of science: That was Kuhn's declared intent, and a project shared by many of his successors (e.g., the SSK movement). Others do aim to identify flaws in science as it is practiced, although their ultimate goal is constructive—they hope to make a good thing even better. Finally, some contemporary people feel alienated from science, viewing it as an institution that threatens important human goals and values.

Why is it so difficult to keep these types of criticism separate? Partly because of the prevalence of the thought that there is a single ideal method that is realized in scientific work—even though as Chapter 2 explained, efforts to say in detail what that ideal might be have met with recurrent difficulties. People who are nonetheless gripped by that thought often suppose that challenges to a philosophical image of science are attacks on the scientific enterprise itself: hence the vehement denunciation of historical and sociological studies of science, from Kuhn on. If science is understood as under siege from an untidy collection of irrational or antirational protesters, those who present a novel picture of scientific work, or who diagnose particular types of biases infecting some areas of research with the aim of overcoming them, will be lumped together with critics who are thoroughly opposed to the broad authority of science.

Diversity and the Feminist Critique

The population of scientists is not representative of the population at large. The stereotypical image of a scientist is of a fairly well-to-do white man, and until very recently this stereotype was substantially correct. Important shifts have taken place in some fields—the rapid increase in the numbers of women in biology and the social sciences is an especially salient change. But in many areas of science this familiar demographic skew remains almost unchanged, especially at the higher ranks of seniority and authority. Does it matter who the makers of scientific knowledge are?

There is reason to think it does. Scientists' choices about what to investigate, and what methods and standards to use, are shaped by the particular constellations of values, concepts, and background beliefs they bring to their work. More recent research in cognitive psychology (see Chapter 3) strengthens the case, showing that human thinking in general is guided by implicit assumptions and mental models learned from early childhood experience and social interaction. These cognitive tools help us to negotiate a complex and fast-moving world, but they are not easily accessible to conscious evaluation or revision. Because people with different kinds of life histories may have different values, make different assumptions, and employ different mental models, they may make importantly different kinds of scientific judgments.

Feminist scientists and philosophers argue that the relative lack of diversity among scientists has contributed to a range of consequential biases in the natural and social sciences. One kind of bias affects choices about what to study, favoring scientific research aimed at producing applications serving the interests of dominant social groups. Thus medical research is disproportionately dedicated to ailments troublesome to wealthy populations, and especially to men in those populations. Vastly more money and time is invested in attempts to tackle relatively minor problems that afflict affluent people than is devoted to many types of infectious diseases that kill and disable children in the poorer regions of the world. Agricultural research, including research on genetic engineering of crop plants, focuses on issues and applications chiefly valuable to the owners of large-scale commercial farms, whereas the needs of subsistence farmers (including many women) are served less well. The feminist critics who have emphasized these and similar points about the distribution of scientific research effort have been very clear that gender bias is only one among a class of distorting factors.

Bias in choosing research topics is apparent in the statistics of funding. Yet the distortions may run deeper. Scientific work that supports prior views about the superiority of currently dominant groups of people may easily be favored, even when substantial evidence is lacking. Historically, research aimed at finding innate biological differences that underlie and explain existing social inequalities has enjoyed intense interest and often won acclaim, even though the details of the explanations that are offered change with the tides of scientific fashion. Bad or sloppy science may be tolerated if it leads to comfortable conclusions. Further, scientists may make uncritical use of concepts carrying substantial background assumptions (e.g., the concepts of gender and race).

An important type of bias results from differences in what scientists who differ in their social position (including gender) tend to notice. In the late twentieth century, a cadre of women entered primatology and began to observe all sorts of things their male colleagues had overlooked. Prior studies of primate troops had often focused on dominance hierarchies, and on aggression and submission among males. Focusing on female choice of sexual partners, the women primatologists were able to document the ways in which subordinate males cultivated friendships with female members of the troop, how they signaled their willingness to help in caring for the young, and how they were reproductively successful in ways the traditional dominance-oriented perspective had never suspected. More generally, the new primatology disclosed a complex web of relationships within and between the sexes, completely revising ideas about primate social behavior (see "A Sea Change in Primatology").

ᑲ A Sea Change in Primatology

From the early decades of the twentieth century on, scientists have observed the behavior of social primates (not only chimpanzees and gorillas, but also baboons and macaques, animals more distantly related to human beings) in an attempt to understand the evolution of behavioral and psychological characteristics in our own species. Until the 1960s and 1970s, the overwhelming majority of the investigators were men. Sometimes they brought back from the field colorful stories of primate life that bolstered the popular image of "man the hunter," but even the most sober and careful researchers focused primarily on social interactions among the male members of the groups they studied. Of particular interest were aggressive encounters among rival males, and the strategies the animals seemed to use to prevent conflicts from escalating. Recognizing the regular tendencies of some animals to give way to others, primatologists mapped dominance hierarchies, and gave special attention to the animals at the top—the "alpha males."

A change of perspective began with the entry of a significant number of women into the field. During the 1970s, women who had already gained some prominence (e.g., Jane Goodall, Dian Fossey, and Alison Jolly) were joined by a new cadre of younger researchers, many of them students of

(cont).

Sherwood Washburn at Berkeley. Like the older scholars, newcomers such as Jeanne Altmann, Barbara Smuts, and Shirley Strum posed a broader set of questions, including within their purview the dynamics of female–female interactions, female–juvenile interactions, and female–male interactions. Their careful studies displaced a number of traditional assumptions.

Primatologists had taken it for granted that high rank in the dominance hierarchy conferred benefits in the only biological currency that matters: Darwinian reproductive success. It was widely assumed that the alpha males were rewarded with a large proportion of the offspring born within the troop (perhaps even all of them), and that females were unable to exercise any choice in mating. Indeed, from the traditional perspective, females were so passive and powerless that they could mostly be ignored—encounters between males were the sites of the real social action. The new cadre of female observers undermined these casual presuppositions, building up overwhelming evidence for far higher rates of mating on the part of subordinate males than had previously been envisaged. That conclusion was the prelude to even more interesting discoveries. Females were recognized as exercising choice about their mates, even at the crucial times (when they are fertile), and subordinate males were shown to act in ways that won the "friendship" of particular females. Primate social life emerged as less simple, less a matter of chest-pounding displays and exercises of strength, and more subtle and intelligent.

To see how far the new observations lead, consider just one among many striking offshoots. Sometimes a female and a subordinate male friend signal to one another their readiness to copulate. Unfortunately, a dominant male is in the vicinity. What is to be done? If there is a useful obstacle nearby—a large rock or a dense thicket, say—it can serve as a screen behind which the would-be mates can retreat. Careful studies have shown that animals can identify places where the gaze of a threatening superior would be blocked. Primates seem to be capable of something akin to primitive "mind reading."

Perhaps the wonderful flowering of primatology would eventually have proceeded even without the women who posed the broader questions, but there are reasons for doubt. Historical and sociological studies of research on the behavior of animals who might serve as models for our ancestors (and for the inherited "springs" of our own behavior) is inevitably pervaded by assumptions and images drawn from familiar features of the observer's culture: Some language must be used to classify what the animals do, and it is impossible to remove all traces of the concepts applied to human contexts. Those who experience recurrent human situations differently can sometimes reveal and displace a perspective that has arbitrarily been imposed on the animals under study. That is what the pioneering female primatologists did, and the field is enormously richer and deeper for it.

In the absence of diversity, research may be biased by casual acceptance of unjustified assumptions, and consequently blinkered. Moreover, as some scholars have suggested, the effect may not be mere happenstance. When women are underrepresented in scientific inquiry, investigations may be pervaded by a general style of thought that is stereotypically masculine. This style of thought emphasizes analysis, abstraction, and appeal to quantitative general laws as the principal means of understanding. It sees causal systems as mechanical combinations of independent atomic components, and aims above all to give us the ability to control the features of the world that matter to us. An alternative style of thought, critics suggest, would use the tools of qualitative thought, metaphor, and narrative; recognize the importance of complexity and context; and seek a kind of personal understanding (especially in the study of human beings and other organisms) rather than simple control (see "The Dynamic Genome: Mobility and Control").

ᴄ⌒∽ The Dynamic Genome: Mobility and Control

Eight years after the rediscovery of Mendel's ideas about inheritance, in 1908, Thomas Hunt Morgan and the "fly boys" of his Columbia lab began a program of research that would transform the study of genes—the newly named heritable units located on the chromosomes—using fruit flies (*Drosophila melanogaster*). They rapidly found a substantial array of easily identifiable mutations, beginning with Morgan's famous white-eyed variant. Tracking the linkages among these mutant genes through generations of crosses allowed the Morgan group to map the genes' locations on fly chromosomes, and to show that they were arranged in order along the chromosomes like beads on a string. Fundamental to the technique of gene mapping was the recognition (the work of Sutton, Boveri, and Bateson) of gene linkage. In the first few years of the twentieth century, investigators had discovered that pairs of chromosomes could "cross over" and exchange segments in the course of egg and sperm production, but otherwise the order of the genes seemed to be stable.

Barbara McClintock began to study genetics only a dozen years after Morgan had begun his seminal work. She took up the challenge of extending the techniques Morgan had pioneered to an organism with a far more complex genetic system than that of *Drosophila*: maize or Indian corn. Maize offered special advantages: The color patterns and other characteristics of the individual kernels of a cob of corn provide a window into the genetic endowment of the embryo it contains, and (once McClintock had discovered how to prepare them for microscopic viewing) the large chromosomes of maize

(cont).

revealed visible structures that made mapping them much easier. Even as a student, McClintock proved able to obtain extraordinary results with the microscope, and she soon developed techniques that allowed her to explore the chromosomes in detail. In 1931 she and her student Harriet Creighton were able to observe chromosomal crossing over.

McClintock turned her knowledge of maize chromosomes to explore a set of mutations created by exposing the chromosomes to X-rays. Among these were some mutations that showed a special property: In the course of development they produced distinctive "variegated" patterns of streaks or spots of color in the plants' kernels and leaves. McClintock hypothesized, and soon confirmed, that the X-rays caused chromosomes to break, and that the ends thus broken off sometimes joined up to form a ring. These ring-chromosomes tended to be lost during cell division, resulting in the variegated pattern. Some variegations proved to have a more complex structure, however. McClintock found that these were caused by what she called "controlling elements"—factors that could move about the chromosome, and that modified the expression of adjacent genes in the locations at which they appeared. This discovery was revolutionary in two ways. First, it revealed the existence of chromosome components (later dubbed "transposons" or "jumping genes") that are mobile, rather than having permanent addresses on the chromosomes. Second, it showed that the chromosomes don't just provide information to be "read off" in the development of the organism, but have components that modify or control the functioning of their genes, perhaps enabling the genome as a whole to respond actively to the environment in which the organism found itself.

Barbara McClintock was awarded a Nobel Prize in 1983 for her discovery of transposition over three decades earlier. Her own theory of genomic control by means of mobile elements was never accepted, but her revolutionary conception of the genome as a dynamic, responsive, and elaborately self-regulating system has proven prescient indeed.

McClintock seems to have been able to achieve results that others could not, by means that she could not always articulate. In comparison to the fly geneticists, and even more to the geneticists who took the lead in the 1950s, working on viruses and bacteria to explore genes' molecular structure, she worked on an organism of remarkable genetic complexity. Her corn stocks themselves were exceptional, capturing tremendous genetic diversity and complexity where other maize geneticists worked with homogeneous inbred stocks. She was able to track genetic interactions in this context by means of an intense and fine-grained familiarity with her plants and her exceptional skill with the microscope: In interviews late in life she spoke compellingly about the importance of having "a feeling for the organism"; of knowing "the whole story" of each plant, from pollination to maturity. Her original

(cont).

insights—ideas like that of the ring chromosome and of the genome as sensitively responding to its environment—seemed to be the result of a special kind of thought process, what some people called "intuition" and McClintock herself called "integration."

Was McClintock's way of doing genetics somehow distinctively a "woman's way of knowing"? This has become a much-debated question. McClintock herself seems to have thought not, but she did think it was something unusual and special to her. One biographer has emphasized McClintock's overriding concern with freedom, and it is easy to see how that might have been an important issue for a woman of her generation (her mother thought she should not go to college, for example, because it might make her less marriageable). Was her focus on the genome's autonomous activity an expression of this concern with autonomy in her own life? Cognitive psychologists have found some evidence that women—at present and on average—are more likely than men to look for holistic and interactive causes, and less likely to seek understanding in reductive simplification. Does this difference help explain Barbara McClintock's insights? Does gender discrimination help explain why they were not more quickly and more widely taken up? These questions remain intriguing and difficult to resolve.

Yet now a puzzle arises. The criticisms we have been considering suppose that individual investigators are likely to frame their research and adjudicate their data in light of pervasive conceptions and values. In this sense, all significant knowledge is biased. Yet the feminist scholars who accept this Kuhnian conclusion are also very critical of certain biases in scientific work. If bias is unavoidable, how can it be a fair target of criticism? And what should we do about it? Are some biases worse than others? Might some even be good?

These questions become less perplexing if we leave the metaphor of bias, and take up instead the metaphor of knowers' perspectives or situations. Knowledge is *situated*: It always belongs to a particular knower or knowers, each of whom occupies a particular social position and has particular background beliefs, values, and interests, as well as a particular set of bodily capacities and skills. But according to a tradition running from the Scientific Revolution to today's science and philosophy of science, the aim of science is precisely to achieve a kind of knowledge liberated from such local roots—a view of the world on its own terms, independent of any particular perspective, a "view from nowhere." The first step to this goal was to start from the best possible perspective. The seventeenth-century founders of the Royal Society believed that *gentlemen*, freed by their wealth and education from distracting personal interests, enjoyed a uniquely clear and trustworthy view of natural phenomena (women were not admitted as Fellows until the mid-twentieth century). The gentlemen scientists' view, further refined by careful adherence to rules of method intended to enhance the objectivity and quantitative precision of their observations and the rationality of their inferences, might (it was hoped) approximate to the perspective-free view to which science aspires. These ideas are plainly products of a particular social context, now long gone. It is time to give up the belief in wholesale perspective-free knowledge, and we should surely also dispense with the idea that social status brings with it a matching epistemic privilege. But none of this requires that we give up the idea that some perspectives might be better—in certain respects—than others.

Some feminist epistemologists have responded to the traditional view just sketched with one diametrically opposed to it, arguing that people in subjugated social positions have distinctive epistemic advantages. Those in socially dominant positions know only their own point of view, the argument goes; the world as seen by the people who grow their food and clean their floors is closed to them. People in subjugated positions, however, must cultivate familiarity with the point of view of the powerful, yet they see it against the background of their own, quite different, perspective. Their situation thus gives them access to multiple perspectives, and affords a special kind of critical distance on them as well.

Some perspectives surely do give special ease of access to particular kinds of knowledge. Any particular constellation of beliefs, skills, aims, and values will make some patterns jump out in the world of experience and obscure others. Are some perspectives superior across the board in the view that they afford? Any conclusion of that sort would be very hard to justify, but the idea that we can identify certain standpoints as offering particular epistemic advantages, either because they encourage careful attention to the application of their own standards, or because they incorporate parts of other perspectives, is a plausible and hopeful one. Perhaps being able to draw on a diversity of frameworks might enable a scientist to uncover lurking biases and come to terms with them. Even more obviously, the clash of perspectives found within a scientific community with many diverse points of view might reveal—to everyone—points at which assumptions were taken for granted, and thus contribute to an improved program of research. If that is right, the inclusion of people of very different backgrounds in the various projects of scientific inquiry is not simply a matter of social justice (opening doors for all), but also a policy that leads to better science.

The Cultural Critique

Science comes from somewhere in the simple sense of having taken shape in a particular part of the world and a particular social and historical context. The contributors to the Scientific Revolution lived and worked in Europe, and the rapid growth of science in the nineteenth and early twentieth centuries took place there and in the growing centers of European emigrant population in North America. Today, every country has its own scientists and its own research programs. Yet the most important institutions for research and training are still to be found in Europe and North America, as are the most influential journals and professional organizations. For many of the world's people, science is still clearly identifiable as something foreign, something that comes from somewhere else. Does the *source* of scientific ideas and practices matter?

If science has special standing as a universally valid method for obtaining objective knowledge, its source is surely irrelevant. But if it is just one local tradition among many, the answer is not so simple. Most scientists and philosophers of science take it for granted that science is quite different from the traditional beliefs and practices that preceded it and that continue to coexist (and sometimes conflict) with it all over the world. Science (it is supposed) is universal, rational, responsive to evidence, and massively successful in guiding prediction and intervention. Tradition is local, often irrational, and dogmatic; its sporadic practical successes are offset by a weight of useless error. But this stereotyped contrast is overdrawn. We have seen that reasoning in science is far more complex, far less algorithmic, and more nuanced in judgment than this idealized picture indicates; further, some very important scientific theories are unable to offer reliable predictions or practical applications. On the other side of the balance, the idea that nobody knew anything before the growth of modern science in the seventeenth century is laughable. Many societies, past and present, have assembled extensive bodies of empirically informed knowledge that are unconnected with the particular intellectual lineage of the Scientific Revolution. Some of these local knowledge traditions, such as the astronomical systems of the Kerala school in India in the fifteenth and sixteenth centuries, have employed sophisticated mathematical systems of representation and calculation. Some have been linked with impressive technologies or other practical achievements, from Chinese explosives and clockwork mechanisms (notably Su Sung's Cosmic Engine, a complex clockwork armillary sphere built in the eleventh century), to Indian medicine (including cataract surgery and a form of inoculation for smallpox, both by the ninth century), to the spectacular feats of navigation achieved by Polynesian and Micronesian voyagers. Some, such as the agricultural practices of farmers in Africa, have been produced by systematic "folk experimentation." Many have been attuned to the local conditions, needs, and values of their users in a way that imported science cannot easily match: Scientists who carry out their fieldwork in regions where "primitive" people live are often struck by how much those people know about local species, about the distributions of minerals, about regularities of temperature and rainfall, and so forth. That should be hardly surprising to anyone who accepts natural reasonableness: When certain sorts of facts matter to intelligent people, you can expect that there will be resourceful efforts to discover them.

Natural reasonableness encourages a more egalitarian view of the relationship between science and other knowledge traditions. Consequently, the question about the source of the scientific tradition takes on a new importance. Western science is displacing older knowledge traditions all over the world. Is this necessarily a good thing?

Some critics have recently argued that the answer is "No." In fact, they say, when science and science-based technologies are brought to societies with little previous exposure to them, the effect can be a failure (witness the attempt to improve on the irrigation practices of the Balinese), or even a destructive failure (new crops turn out to have unanticipated environmental consequences, effects that the traditional practices had learned to avoid). Sometimes the innovations are a means of cultural domination and economic exploitation. Accusations of this sort shock many who hold an idealized view of science. But a look at the role that science has actually played in shaping the global political and economic landscape will provide grounds for taking the criticisms seriously.

The history of science is closely intertwined with the history of European colonial expansion. The voyages of discovery that are celebrated for their place in the history of inquiry were, of course, mainly aimed at finding valuable natural resources and routes for transporting them back to Europe. Scientists and engineers contributed to the colonial project in many capacities—developing new instruments of navigation and better weapons, identifying and mapping biological and geological resources, and helping to develop new ways of extracting and using the resources the new colonies supplied. Scientific ideas were also used to lend moral legitimacy to the treatment of indigenous people under colonial rule. Theories of race represented many non-Europeans as unfit to rule themselves, sometimes as less than fully human, and colonialists also saw their own new technologies as compelling evidence of their superiority to the people they enslaved or displaced. Science thus played a crucial and highly visible role in the creation of Europe's colonial empires, with their distinctive structure of power relationships between the colonizers and the colonized. For people who think they were harmed by those relationships, the connection can seem a damning one.

Although the colonial empires are now almost gone, critics see science as continuing to play the same old roles in new contexts, aiding the powerful in ways that are often harmful to the relatively powerless. Scientists continue to provide justification for global inequalities in the form of (purported) biological explanations of disparities in income or educational achievement. Most scientists and engineers are now employed by corporations and military organizations; their work perforce often serves the interests and values of these employers. (One well-known example of the use of the tools and authority of science to forward corporate interests with little regard for their human cost has been the promotion, in hospitals, of "scientifically designed" infant formula to mothers who have no access to clean water.) Even when scientifically informed interventions are undertaken with benign intentions—introducing new hybrid crop varieties with higher yields, for example, or building dams to provide hydroelectric power—they often involve little or no consultation with the people whose lives will be most directly affected—with farmers who will be obliged to buy expensive seed every year rather than saving their own, or people who will be displaced by the dam's waters.

A broader failure to engage in serious two-way communication infects the interactions between science and traditional knowledge systems. Science education, and training in science-based practices through agricultural outreach or public health education, is typically a one-way communication. Teachers engage little with the background beliefs of their students; their aim is to replace these, not to understand them. Scientists or technicians instruct farmers about how to improve their practices to succeed with new crop varieties, or teach people to use mosquito nets to protect themselves from malaria. But they often make little effort to learn about how the people they are instructing understand their own situation and the needs that the intervention is meant to remedy. If the new crop varieties require water that cannot be reliably supplied, or if the mosquito nets are cumbersome to use for families sleeping together in a single room and make little sense in light of local beliefs about disease, the interventions will fail unless their authors elicit that information from the people they set out to instruct. This sort of asymmetry reaches an extreme in the sharp divide between the agents in research programs—the scientists conducting the inquiry, and the employers they serve—and people whose role is more object-like, including technicians or informants who serve an instrumental role in carrying out others' agendas, and those who appear simply as objects of study.

Scientists do sometimes find there is valuable information to be gleaned from local knowledge traditions. International legal disputes concerning the practice of "bio-prospecting" or "bio-piracy" have drawn attention to the self-serving epistemic double standard that can result. Scientists investigate the medical or agricultural practices of a particular local tradition, and obtain samples of medicinal plants or crop varieties in local use. They (or the companies that employ them) then claim intellectual property rights as the discoverers of the crucial chemical compounds or genetic information, sometimes seeking to control access to this proprietary information by the very people from whom they obtained it. Bio-prospectors argue that this practice is legitimate because the local tradition is not the result of scientific research by identifiable individuals: It is part of nature, not part of science. Critics regard this argument as both dishonest and exploitative, a clear indication that what distinguishes "science" as such is just whose interests it serves.

On the face of it, most of the concerns raised here seem not to be about science itself, but about how (and by whom) it happens to have been used. To respond to these concerns, we would need to find ways to enable a more diverse population of scientists to participate fully and creatively in the scientific enterprise, and to ensure that the sciences are answerable to the needs and perspectives of people from a broad spectrum of social and geographical situations. A more fundamental worry, however, seems to lurk behind some of the concerns noted here. Is there something about science itself that makes it especially apt to uses that reinforce existing social inequalities, or that cause it to support certain social values and undermine others? This is a serious question, not one about which we should rush to judgment. We return to consider it directly in Chapter 6.

The Ecological Critique

Many cultural critics would take a further step, suggesting that the local knowledge that is often discarded or displaced in the march of science has a very different character from the scientific knowledge that succeeds it. It is more attuned to details, more concerned with fitting details into a system, more "holistic"; by the same token, it is less "analytic," less concerned to force a range of phenomena into the purview of a simple model. Judgments like these inspire a related, but distinct, form of critique.

When mechanical philosophers described organisms as "machines," one key implication was that we could explain an organism's behavior by looking at what its component parts are like and how they interact, just as we would look at the wheels and springs of a clockwork mechanism to explain how the clock's hands move. This analytic method—understanding the whole in terms of its parts—has proven to be a spectacularly powerful one, and is ubiquitous in the natural and social sciences today. But perhaps the method's undeniable successes lead scientists to overlook its limitations.

Sometimes the behavior of a whole is best explained in terms of its properties *as* a whole, not the properties of its parts (see Chapter 3). This is especially true for dynamic, tightly integrated complex wholes such as cells and organisms. A scientific approach that looks only at the parts of an organism (e.g., its genes or its brain chemistry) is likely to make serious errors just because it does not take adequate account of the complex interactions among the parts that produce the functional properties of the organism. The difficulty of grasping all those complex interactions lies behind the well-known difficulties of medicine and agriculture: Understanding of the parts translates only unevenly into successful interventions—what works in the lab, or in vitro, breaks down on the farm, or in vivo. The most impressive successes come when a single causal factor is responsible for disease, and our understanding of that factor enables researchers to intervene in the process—developing antibiotics for bacterial infections, vaccines for preventing other infectious diseases, and insulin treatment for diabetes. When the causal background is more complex, medical progress goes more slowly. Surgical techniques to repair faulty "parts" such as damaged hearts or clogged arteries have advanced dramatically, but our ability to prevent heart disease and atherosclerosis has not kept pace with these advances. Although we have learned a great deal about many of the relevant causal factors, we do not yet understand how those factors interact to produce the organism-level disorders we wish to overcome.

The analytic approach sometimes inclines investigators to pay too much attention to what is going on inside an organism. Researchers lose sight of the outside influences—the external environment and relations with other organisms. Critics argue that the focus on genetic research in the "war on cancer," and the relative neglect of environmental causes other than tobacco, is an example of this kind of error. Similarly, in agricultural research, scientists working on finding chemical means of killing pests and weeds failed to anticipate either the wider effects that pesticides and herbicides would have as they moved through the food chain, or the rapid evolution of resistance in weed and pest species.

An analytic approach to knowledge can also induce disregard for the social, political, and economic contexts of the phenomena we are interested in, contexts that may be crucial to understanding the problems we wish to solve. Consider the aims and effects of the Green Revolution research program. With the aim of supplying adequate food for rapidly growing populations in India and elsewhere, researchers sought to increase agricultural production by modifying the genetic properties of crop seeds. In an obvious way, the program was a signal success, producing hybrid strains of some staple crops that were capable—under optimum conditions—of producing yields far greater than those produced by traditional varieties. Indeed, the total grain production in countries where these new hybrids became widely used did increase impressively. But this approach to increasing the food supply disregarded the complex web of relations connecting the expression of genes in a field of rice or corn with the economic and social lives of farmers and their communities. The effects of the introduction of the new hybrids were often contrary to the researchers' expectations: Although crop yields increased, the well-being of many people—including their access to adequate food—was impaired. How did this happen? The lower food prices provided by the new seeds were beneficial to urban consumers, but the greatest benefits accrued to industrial-scale producers, the purveyors of the expensive hybrid seeds and the fertilizers they required, and businesses trading in the crops they produced. Small-scale farmers often could not afford the costs of irrigation, fertilizer, and the seeds themselves without going into debt, and risked crop failure (and inability to repay their loans) when they could not maintain the special conditions that the hybrids needed. Lower prices only exacerbated these problems. As small farms failed, the people who had depended on them for their subsistence were forced to leave their rural communities, joining the flow of migrants competing for jobs in overcrowded urban centers. The transition to the new varieties also undermined agricultural sustainability: The agricultural practices required by the new hybrids caused soil degradation over the longer term, and genetic resources were lost as traditional crop varieties were abandoned (see "The Transformation of Agriculture").

✍ The Transformation of Agriculture

Although it is usually associated with rice and with Asia, the origins of the Green Revolution lie in wheat in Mexico. In the early 1940s, the Rockefeller Foundation collaborated with the Mexican government on a project to increase Mexican agricultural production, in part by finding a way to reduce the losses of wheat to a fungal disease, wheat rust. One member of the team, a young plant pathologist from Minnesota named Norman Borlaug, began a selective-breeding program to develop rust-resistant wheat. To maximize yield, Borlaug used the irrigation systems and inorganic fertilizers that had become widely used by large landholders in the United States. To speed up the breeding process, he shuttled his wheat stocks between two regions in very different climatic zones so that he could raise two crops a year. This had an unanticipated effect: His new strains of wheat could succeed in many different environments, as long as irrigation and fertilizer were provided.

Within ten years Borlaug had produced strains of wheat that were resistant to many common diseases, and that responded to irrigation and fertilization with yields close to double those of traditional varieties. A further breakthrough came when he crossed one of his successful new varieties with a Japanese semidwarf strain, enabling a threefold increase in production.

Borlaug's new wheat cultivars were taken up rapidly by farmers who could afford the irrigation equipment and fertilizer that was needed. By 1963, 95 percent of Mexico's wheat crop was made up of Borlaug's cultivars, and the crop as a whole was six times what it had been two decades before. Some of the increase resulted simply from improved yields per hectare. But some resulted from the displacement of the maize and beans traditionally raised for local consumption, in favor of wheat destined for urban or international

(cont).

markets, and by the expansion of agricultural production into new land as a new class of entrepreneurial farmers were granted generous government subsidies to irrigate land too dry for traditional agricultural use. The poorest farmers in Mexico—including many women and indigenous people—were unable to compete in this new agricultural economy. Many were forced to sell their land, and because the new agriculture had replaced human labor with machinery wherever it could economically do so, there were few jobs for landless farm workers. Most of the dispossessed moved to the rapidly growing urban areas in search of factory work.

In India, meanwhile, drought and internal struggles over agricultural policy led to repeated food crises in the 1960s. A geneticist, M. S. Swaminathan, recognized that Borlaug's new wheat strains had the potential to increase India's food production. In 1965 the Indian government agreed to collaborate with the Ford Foundation and the Rockefeller Foundation in large-scale planting of seed purchased from Borlaug's Mexican stocks, and to launch an aggressive program of investment in irrigation infrastructure, fertilizer production, and government assistance to farmers to support the transition to the new agriculture. The success of this crash program encouraged attempts to extend the methods that had been so successful with wheat to rice breeding. The Borlaug recipe led to rapid success, and the new varieties and agricultural methods increased India's grain production rapidly. By 1970, when Norman Borlaug received the Nobel Peace Prize for his work in increasing the world's food supply, India was self-sufficient in grain production; the total rice yield had increased by a factor of about six, and the Green Revolution was well underway.

As in Mexico, India's increase in total yield came with some costs that were not recognized as widely or quickly. The high capital costs of the new agriculture—which required not just seed and fertilizer, but expensive equipment, fuel, and pesticides to protect the crops against insects and weeds that traditional agricultural practices had controlled by crop rotation—drove many poor farmers into debt from which they could not escape. Others were unable to afford the technological resources the crops required, and ended up with yields far below what traditional varieties would have produced. Many smallholders lost their land. Ecological effects began to become obvious, as soils became waterlogged or salinized by irrigation, and the plants and fish that had lived in the margins of rice paddies, providing important additional sources of food for farmers, were killed off by pesticides. And a quiet loss of genetic resources began: The thousands of diverse varieties of wheat and rice that had been developed by farmers in the Indian subcontinent over millennia, many with special adaptations to particular human needs or environmental conditions, were abandoned and began to be lost. Finally, in India as

(cont).

in Mexico, the success of the Green Revolution brought permanent changes in many people's ways of life, undermining the economic viability of villages supported by small farms, increasing migration into the new industrial economies of the urban centers, and reducing women's economic independence and role in community leadership.

The breeding programs and agricultural interventions of the Green Revolution had been chosen by leaders of governments, by officers of the Ford and Rockefeller Foundations, and by scientists with backgrounds quite remote from those who were supposed to benefit from the agricultural innovations. The concerns of the decision makers were different from those of the farmers. The U.S. government and U.S. foundations shared an interest in reducing the threat of political instability: They worried about the effects of population growth outstripping food production. The Green Revolution was thus seen as a means of preventing a Red Revolution—the spread of communism. The Mexican and Indian governments were both at sensitive points in postcolonial consolidation, in which they saw a choice between further land reform to support an egalitarian agrarian economy, or a path to increased modernization, industrialization, and urbanization. Internal pressures in favor of the second path coincided with the offers coming from the United States. The scientists were concerned—often passionately concerned—to relieve hunger, but their understanding of the causes of hunger was very simple. For them, the question was always how to produce more food. They did not appreciate the much more complex and difficult problems of its distribution, and overlooked crucial differences in the circumstances of agricultural production. What could have been achieved by a comparable investment in research and implementation of agricultural changes chosen in consultation with a more diverse group of stakeholders, including farmers and village leaders? It is difficult even to guess.

The point is not that science, pursuing an analytic approach, is unable to uncover the structure of the complex webs of causal relationships among the parts of an organism, or between an organism and the various elements of its environment, for of course all these relationships can themselves be investigated analytically. The worry, rather, is that science in the analytic mode tends systematically to neglect these relationships, and that this myopia can lead to a distorted representation of important aspects of the causal structure of the world and thus to interventions that "bite back," producing unexpected and undesirable consequences. The ecological critique can be understood as a constant warning against oversimplifications that are often introduced to ease the work of analysis.

Antiscience

The critics we've been considering are sometimes accused of being "antiscience." That charge is unfair. For the critics take very seriously an idea celebrated by the most ardent defenders of science: Scientific knowledge is potentially capable of dramatically improving the human condition. Comparing that idea with particular historical and contemporary practices, they urge changes that might promote that worthy goal. A greater diversity among investigators would help expose and eradicate biases; attention to the situations and understandings of people whose views are often dismissed as primitive, and awareness of the possibility of oversimplified analyses, would guard against scientific "solutions" with counterproductive consequences. The critics want to make science better.

Nevertheless, the world clearly contains genuinely antiscience sentiment: attitudes of hostility or contempt for science that have important effects on public decision making about science policy and hobble our ability to make good use of scientific knowledge. Sometimes the roots of resistance to science lie in the considerations that underlie the critiques reviewed in previous sections. Colonized or marginalized people who have suffered the adverse effects of incautious or hostile applications find science alienating. Yet these are not the only sources of concern. Worries about the human implications of the scientific worldview and the scientific conception of knowledge go back to the early modern period when physics achieved its earliest successes.

The triumph of the "mechanical philosophy" in the eighteenth century provoked worries that science had given us a world that is soulless, meaningless, and devoid of all the things that matter to people. Before the Scientific Revolution, Europeans lived in a world that they understood to be filled with meaning and purpose, one in which the place of humankind was (in every sense) central. They were "at home" in the universe. The cosmos as they conceived it was of a reassuringly human scale, and its history stretched back only a few hundred human generations. Their knowledge about the world—about its structure and history—was at the same time knowledge about the meanings and purposes of their lives and actions, knowledge about how to live. Other traditional systems of belief have shared many of these features, although, of course, the details have varied widely. But the world revealed by the new sciences of the seventeenth century was quite different. The Earth was no longer the center of the universe, but just one planet among many. The universe was vast and impersonal, and devoid of the qualities most familiar to us; the homely world of our sensory and emotional experience came to be seen as an illusion laid over an austere reality—atoms moving through the void. Knowledge, as the new sciences came to conceive it, is a dispassionate knowledge of the facts of this austere world, and nothing more. It cannot help us to understand our place in the universe, the meaning or purpose of our actions or of larger events. Indeed, as the new sciences matured, they seemed more and more strongly to suggest that there are no meanings or purposes at all, except those that we invent; no truths about what we should do or how we should live, beyond our conflicting desires and opinions. In an image repeated again and again in writings by scientists and philosophers, we heirs of the Scientific Revolution found ourselves alone, in a "disenchanted" world, a world that cares nothing for us, a world unimaginably vast, empty, and cold.

The scientific developments of the ensuing centuries offer no new comfort. Quantum physics has given us a world that is far stranger and less comprehensible than that offered by the mechanical philosophy, but no more humane. And most important, evolutionary biology, cognitive science, and neuroscience have combined to bring the full force of the scientific worldview to bear on our understanding of ourselves. Descartes and the other mechanical philosophers abolished the idea of organisms as beings endowed with their own purposes and meaningful structure, but though they saw other organisms and human bodies as elaborate machines, they held that the machine that is a human body is connected to a soul. The new sciences of our era turn even the soul into part of the machine.

Defenders of science have little patience with these complaints. We should stand firm, they say, even if the truth is hard to face. If we give up the cozy illusions of the past, we may come to a different kind of fulfilling experience, one reported by scientists from Galileo to Richard Dawkins, achieving the thrilling sense of intellectual freedom science offers and appreciating the awe-inspiring beauty, order, and richness of the world it reveals. A clear-eyed vision of reality more than compensates us for the loss of our prescientific illusions. We should also remember that the sense of meaning and purpose inherent in the worldview of prescientific Europe was deeply involved in the justification of a social order marked by an oppressive class structure and dogmatic religious belief. Liberating people from the "purposes" of such a social order was a great achievement, one made possible by the emergence of modern science.

Many people are not convinced by defenses of science along these lines. They have a deep sense of loss, one not to be assuaged by urging them to share the thrill of understanding nature without illusions. Some urge a response along lines already noted in connection with the critical perspectives we have looked at—the development of a scientific perspective that resists reductionism and abstract thinking in favor of a style of thought sensitive to the meanings and purposes that people find in nature (this approach has a long history, with roots in the Romantic science pioneered by Goethe in the eighteenth century and influential in the emerging life sciences in the nineteenth century). But most take a different path. Although they may respect areas of science bountiful in offering predictions and interventions that make their lives go better, they seek ways of interpreting the theoretical bases for the practical successes that allow them to retain comforting beliefs: If the cosmologists tell them that the universe originated in a "big bang," they continue to envisage a creator who lit the fuse. And where predictions and interventions are less successful, they are apt to be skeptical of scientists' claims. As they focus attention on human sciences, they recognize forecasting failures and misguided interventions—the economists are far less sure-footed than their counterparts in chemistry. Especially with those sciences that deal with complex systems—historical sciences like evolutionary biology as well as studies of ecosystems and of the Earth's atmosphere, where precise predictions and impressive interventions are hard to come by—there might seem to be grounds for doubt. If the standard of success is prediction and control, then many alleged sciences appear to fail the test. Especially when these sciences clash sharply with firmly held views about the place of humanity in the cosmos, they are apt to face firm rejection.

Public disagreement among scientists exacerbates the situation, perhaps because many people have expected that science would offer certainty. Even when, as in the case of climate change, thousands of experts agree, the force of the consensus can be blunted when people perceive that "other experts" (typically not specialists in climate science; typically with well-concealed sources of support from industrial conglomerates) offer an "alternative viewpoint." It is not hard to hang on to your cherished convictions when the science you are asked to accept has little to offer in the way of spectacular interventions or precise predictions, and when the debates carried on in the media seem to make it clear that there are two sides to the story.

Antiscience is not exactly wishful thinking, but it grows out of a desire to preserve what the sciences seem to threaten. Its growth is facilitated when the most striking markers of scientific success are absent, and when the clashing "experts" seem to show that the issue remains controversial. Instead of thinking of antiscience sentiments as wholly irrational, we might view them as expressions of natural reasonableness in a social environment that hinders the transmission of well-supported findings. Central aspects of people's ways of life are threatened—they are asked to give up their normal practices (e.g., of energy consumption), their commonsense beliefs about the world, or (even worse) the deity in whom they put their trust. The exhortations to find fulfillment in the bracing experience of looking truth in the face make little contact with the realities of most human lives. The evidence that supports the threatening sciences is subtle and delicate, difficult to explain thoroughly and lucidly. And finally, what percolates through the channels of information transmission is not even an approximation to a simplified version of the evidential situation, but a cacophony of voices in which the experts cannot be distinguished from the ignorant or from those whose verdicts have been bought. No wonder, then, that antiscience is one result.

Science as a Social Endeavor

The story of how people come to feel resistance to science, and become alienated from it, has a further twist. As the public becomes accustomed to think about science as rife with debates, it becomes very clear that scientific work is carried out by people, people who have ambitions, loyalties, values, and personal goals. For those who have been convinced by the ideal of science as value-free, suspicion arises that what is going on is a highly imperfect form of science—and that if the frequency of controversies is so high, perhaps science rarely (or never) lives up to its supposed standards.

The last chapter explained how scientific work is permeated by values, suggesting that we can acknowledge the pervasiveness of values without opening the door to relativism. By the same token, even though the scientists who produce new findings are people with allegiances and personal ambitions, embedded in social networks and broader societies, it does not follow that their work is inevitably corrupted. As we noted earlier in this chapter, critics of an influential image of science are often taken to be antiscience, and that charge is often leveled against historians and sociologists of science. But the common perception that a particular area of controversy is corrupted by the incursion of values that should have no place in scientific debate is not the product of any profound study of Kuhn or of the ideas advanced by SSK. The participants in the debates make the challenges themselves, repeatedly: Opponents of evolution are in the grip of literalist religion, Darwinists have a materialist agenda, deniers of global warming are pawns of oil companies, and climate-change Cassandras are antibusiness tree huggers.

Chapter 6 investigates more thoroughly the consequences of recognizing the features historians and sociologists have discerned in the practice of science. The aim of this section is to scotch the idea that recognizing scientists as human, and as members of human societies, automatically undermines respect for their accomplishments. Although philosophical accounts of science typically ignore the social embedding of scientific practice—conceiving science as something that goes on in the mind of an individual with some consistent set of beliefs, or even a precise assignment of probabilities to hypotheses—recognizing the social involvement of researchers should not inspire the conclusion that attention to the evidence is inevitably overwhelmed by baser urges.

Start from a point that emerged in discussing Kuhn's ideas about revolutionary disputes. When rival paradigms confront one another there is no "instant rationality." Instead, it is reasonable for some to jump onto one bandwagon, and equally reasonable for others to leap a different way. In the sixteenth century, for example, there were many traditional astronomers, a handful of Copernicans, and even more occasional mavericks (like Tycho Brahe), who offered a distinctive novel view. Yet, supposing that it was reasonable to try any of these schemes, things could have worked out differently: All the members of a community of reasonable scholars might have reasonably decided to opt for exactly the same framework. Maybe, after Copernicus (who conveniently died after receiving the first copy of his magnum opus), there would have been no more Copernicans. Surely that would have been a bad outcome for science, far inferior to the actual process of dissension and prolonged debate. A simple point emerges: A community of reasonable people is not necessarily a reasonable community.

What, then, makes for a reasonable community? Premature uniformity seems to be a bad thing. Some dissent is good when the evidential situation is unsettled—when there are alternative frameworks available, frameworks that have different virtues. It may even be good for dissent to persist for a significant period, so that the position that eventually achieves consensus endorsement should truly earn its triumph. Even after the consensus has come, some dissenting voices might still be welcome, offering different perspectives that could open up inquiry in new directions. (Barbara McClintock's work on "mutable loci" on maize chromosomes is a case in point. Her dissent from the Mendelian consensus in genetics pointed the way to a revolutionary rethinking of the nature of the genome.) Dissent can, however, be overdone. Sometimes, because of social pressures, scientists are invited to engage in the same debates again and again, with opponents who feel quite entitled to ignore any evidence that has been presented in previous rounds—as, for example, in the recurrent attacks on evolutionary theory, where critics continue to claim that evolution is incompatible with the second law of thermodynamics, even though the point has been patiently refuted hundreds of times.

Something like consensus is often needed, for societies rely on scientific research in crafting their policies. We can see the dangers of not appreciating a scientific consensus when political pressures artificially manufacture forms of dissent and hide the basic agreement, thereby interfering with policies that are urgently needed: We have only one planet, and feckless, ignorant, and myopically self-interested leaders may have done humanity great harm by failing to attend to what responsible scientists said about it. Celebrating the proliferation of dissent in permanence ignores the importance of responsible decisions and responsible actions based on them. Plainly, then, two extremes are to be avoided. An entirely homogeneous scientific community is a bad thing, and so, equally, is one that never reaches (approximate) consensus on anything. What exactly is the right amount of cognitive variation? And how are we to achieve it?

There is no general answer to this question, for everything depends on the issues under discussion, and on the risks that come with particular kinds of errors and with delay. Yet we can illustrate some important points about diversity of opinion within a scientific community by considering an artificial case, one in which those involved know a lot more than scientists usually do.

Suppose it is important to discover the structure of a particular molecule, and there are two rival views of how to proceed in fathoming it. Everybody thinks (correctly) that one approach is far more likely to be successful than the other, in the sense that, given equal assignments of labor, the chances of success along the first route are greater than those along the second. Nonetheless, the second approach does have some promise. Everybody also knows that if one approach works, its rival is bound to fail. There are ten qualified scientists, all ready and willing to participate. If all ten worked on the first approach, the chances of success would be only slightly greater than what they would be if only eight pursued that approach. One person working on the second approach would offer a small chance of success, and the probability would go up considerably if two were to try this route. Thereafter, additional people working on the second approach would barely boost the probability of its reaching fruition. What is the most reasonable way for the community to proceed?

We have rigged the example, introducing assumptions about the likely effects of dividing cognitive labor that nobody could know in any realistic case. Given these suppositions, the answer is obvious: It is best if eight work on the first method, and two on the second. That would be the preferred strategy. Can it be reached? Not if each of the scientists acts independently in the high-minded fashion of traditional philosophical images of scientists. For independent rational scientists will each suppose that the first method is preferable to the second, and, through their separate decisions, the community will gravitate to a state of homogeneity—better, to be sure, than the other mode of homogeneity (where all pursue the second method), but inferior to a mixture of approaches.

Suppose, however, that we bribe the scientists a bit. We institute a Prize, of great prestige, a noble award to be given for the first successful resolution. We also allow them to know what their colleagues are doing. Considering a situation in which the first eight have signed on to the first approach, newcomer number nine reasons as follows: My chances of getting the prize depend on two things, the promise of the method I choose, and the number of competitors who are using that method; if I join the rest, I shall be committed to the better approach, but I'll only have one chance in nine of getting there first; the probability of winning is better if I take up the second. Number ten reasons similarly, and the community achieves the most reasonable distribution.

Commentators on the social embedding of science sometimes delight in suggesting that the enterprise is distorted by the values of individual scientists, from their grandest political ideals to their most venal proclivities. Philosophers often react by inventing an image of scientists as hyperrational, so devoted to the truth that their "degrees of belief" reflect the evidence and nothing but the evidence (although, if pressed, few would suppose them to be saints in lab coats, freed in some mysterious way from the pressures that afflict and the ambitions that inspire ordinary mortals). Philosophical discussions tend to overlook a point that leaps out from historical and social studies: Scientists are often driven by a desire to achieve recognition—we might even say that they work in a "credit economy." Our toy example is intended to show that extrascientific motivations (the desire for credit, dramatized in the Prize) can actually *promote* good community strategies. Independently acting high-minded scientists would do less well; high-minded researchers who reflected on the distribution of their colleagues' effort, and who were prepared to do whatever was best for the community effort, would do equally well but no better. Those who are driven solely by ambition can be given incentives that will make the community strategy as reasonable as it can be. This is not to say that the credit economy always works smoothly, for competition for credit could have other consequences that are damaging—perhaps researchers will be less disposed to share information or more inclined to cut corners in designing or carrying out experiments. The reasonable coordination of research effort is a matter for philosophical exploration.

Is there enough diversity of opinion and of approach in contemporary research communities? It is hard to tell, but the answer is probably "No." Yet an existing social mechanism surely does promote some cognitive diversity, along just the lines we have indicated. Scientists know very well that one can win high honors, including Nobel Prizes and respectful references in future textbooks, through normal scientific activity. They know also that the really large heroes in the scientific pantheon are the revolutionaries, those who offer a very different framework for a field. Of course, the vast majority of those who offer radical new proposals are dismissed, vanishing from science and its history, sometimes suffering unemployment and penury. Normal science recruits the workers it needs because the probability of success through desertion is typically infinitesimal—better in such cases to play it safe. The calculus of chances alters, however, when the normal scientific tradition uncovers anomalies. At that point, there are risks both ways, and the ambitious and self-confident may choose to cast themselves as revolutionaries. The reward structure of science, where the largest honors go to those who take great risks, thus enables us to understand how, in the early stages of scientific revolutions, novel ideas can attract the adherents they need to give them a chance of survival.

Much more radically "socialized" perspectives on scientific knowledge have been proposed. These have often been viewed as so outlandish, so antiscience, that acrimonious debates—the famous "Science Wars"—have erupted around them. One of the insights many historians and sociologists have drawn from Kuhn is that value judgments permeate the frameworks that scientists adopt (Chapter 4). Among those judgments are commitments both to the relative importance of particular kinds of problems and to standards for evaluating lines of argument. In a series of detailed sociological studies, scholars have argued that these value judgments are themselves shaped by the social conditions under which research is done.

In this spirit, in one of the most influential discussions in recent history and sociology of science, Steven Shapin and Simon Schaffer review the debate between Boyle and Hobbes in the 1660s—a controversy ostensibly focused on the functioning of the air pump and the experiments Boyle performed with it. They contend that this controversy was simultaneously concerned with the standards for empirical inquiry and the proper form of social order. What standards of evidence should be employed in assessing various sorts of observations and experiments? What rules should govern discussions and debates in the nascent Royal Society? How, at a critical moment in British political life, should the fundamental social institutions be reshaped? Although Shapin and Schaffer sometimes veer toward the relativistic positions discussed in Chapter 4, their most central arguments are dedicated to showing the interconnections of questions about values at many different levels, and, particularly, how large questions about how to order British society affected decisions about how to respond to the things that went on in Boyle's pump (see "New Methods of Inquiry and Government").

᥆ New Methods of Inquiry and Government

Robert Boyle, a scion of the Irish aristocracy, was a leader in developing and articulating the new experimental practices of the seventeenth century. In 1657, Boyle and his assistant Robert Hooke began a series of experiments with the newly invented air pump, a complex and expensive device able to pump the air out of a large glass globe within which other apparatus (or other things, such as live animals) could be placed. In 1660—the year of the Restoration of the monarchy after the turmoil of the English Civil Wars and their aftermath—Boyle published his account of this research: *New Experiments Physico-Mechanical, Touching the Spring of the Air, and its Effects.* The next year, the renowned (and much older) political and natural philosopher Thomas Hobbes published a book criticizing Boyle's work. The ensuing dispute continued for over a decade.

(cont).

At stake, on the face of it, was the status of the new experimental philosophy as a way of obtaining knowledge. Hobbes defended the traditional method of combining everyday observation with deduction, under the guidance of Euclid and other ancient authorities. Boyle's new method put aside the authority of the ancients, and looked beyond simple observation to experiments using specialized apparatus—machines like the air pump—to create entirely new kinds of phenomena intended to reveal the underlying structure of the natural world. Hobbes saw natural philosophy as aiming for certainty like that of the theorems of geometry, a goal beyond the reach of the new experimental method, whereas Boyle saw the facts observed in the laboratory as the most secure basis for knowledge, and accepted that such knowledge could only aim to establish its claims as highly probable.

Other experimenters found Boyle's results difficult to reproduce, however, for the apparatus was far from easy to construct and use. Boyle's success in persuading the emerging scientific community therefore depended on another way of "sharing" his laboratory experience. Boyle used a new kind of writing: clear and simple reports—with diagrams—of what went on in the laboratory, with no literary flourishes and no appeals to authority. These reports gave readers the sense of having witnessed the experiments themselves, and set the model that scientific writing has followed ever since.

This means of persuasion worked by creating a new authoritative body: the community of modest, reasonable experimenters who made up the new Royal Society. Boyle suggested that this community was more trustworthy than the top-down authority of monarchy and priesthood, whereas Hobbes argued instead that it functioned as a closed committee whose secret methods were not accessible to the public, and whose challenge to existing authority could only result in hazardous social destabilization. The dispute over the experimental philosophy was thus not just concerned with the justification of claims about the nature of the air and the difference between experiment and everyday observation. The Restoration, with its new experiment in parliamentary monarchy, created a delicate and dangerous political landscape—one fraught with tension among the monarchy, the Church, and aristocracy and commoners in the Houses of Parliament. Both Hobbes and Boyle understood that their disagreement was in large part about who should hold political authority and how best to maintain social order, and these issues played a central (and inescapable) part in its negotiation and resolution.

Debate within "science studies" is often vitiated by a tendency to attribute the most bizarre views to the opposition. Many philosophical critics have thus been distracted from an interesting, and possibly correct, sociological thesis. Part of what may be at stake between disputing scientists is a disagreement about what goals are worth pursuing in research. On the one hand, scientists' conceptions of worthwhile goals can depend on their larger views about human beings, human needs, and the modes of social organization appropriate to those needs. On the other hand, choosing particular goals has consequences for the methods you take to be useful, in that warranted methods are those likely to prove successful in attaining the goals. Seen in this way, the substantive standards of scientific inquiry are quite reasonably interwoven with broader social concerns.

Appreciating this possibility allows for a more sympathetic view of the notorious idea of *social construction*. One interpretation of the thesis that the world (or the truth about it) is a social construct would conjure up a collective mind that, in some mysterious way, brings nature into being. Traditional forms of idealism sometimes propose that the world is shaped, or even constituted, by the activity of the knowing subject. Read in this way, social constructivism would be an odd development of idealism, one that invents an implausible collective entity to do the shaping or constituting. A more sober approach can accommodate a robust sense of reality that rejects idealist fantasies. In all sorts of prosaic ways, the world in which we live is stamped with previous social decisions. Because people have collectively decided to pursue particular projects in science and technology, the landscapes in which we live have been greatly altered. Even the "natural" parts of our world, the trees and grasses and flowers around us, often are as they are because of extensive programs of planting and clearing, breeding and selection, guided by socially validated human desires and by scientific research carried out to satisfy those desires. The world we inhabit is also shaped in more subtle ways by the classifications through which we view it, the connections we make in our thoughts. We experience it as divided into types of things, where the concepts of the types have been fashioned to enable us to pursue inquiries that realize our goals. In these evident senses, the world is socially constructed.

Consider in this light one of the examples with which we began (Chapter 1). In mid-twentieth-century America, people lived in a society divided by race. Today, many anthropologists and social theorists declare that race is a social construct. What exactly does that mean? Not that there are no differences among people—for there are. Extensive recent research on the genetic structures of human populations has shown that there are statistical patterns in the distribution of certain alleles among human groups that represent, unsurprisingly, past patterns of migration, isolation, and inbreeding. The concept of race was born from the tacit recognition of patterns of heritable difference among groups of people who came from different parts of the world, people who had long been separated from one another, and who were now beginning to meet. The decision to treat those differences as important was a social one. Nature did not shout out to our predecessors that this was a division worth marking—indeed, it turns out, the vast majority of human genetic variation is found within groups rather than between groups. Instead, they decided, largely for reasons we now find repellent, that the distinctions among the various inbred groups were significant enough to be explicitly marked, and that the marking should be accompanied by a further apparatus that made judgments of superiority and inferiority. There were biological realities behind the racialized world of Jim Crow America, but that racial world was also a social construct (see "The Biology of Race" in Chapter 2).

The story of how the sciences responded—and contributed—to the construction of the racial world is a complex one. On the one hand it includes the long tradition of racialized science with its claims about racial differences in intelligence and social behavior. On the other, it includes work that helped challenge the preconceptions of Jim Crow America, and contributed to its (perhaps partial) dismantling. This history reveals how science can reflect and reinforce existing social inequalities and cultural biases, as feminist and cultural critics have emphasized. It also shows something about how the kind of self-correction noted in Chapter 4 can come about. Both possibilities are rooted in the inescapable necessity of selecting certain features of the world as the important ones.

The history of the sciences is a history of such selections. At different stages individuals and communities decided what problems were worth pursuing, and, in so doing, they responded to the antecedent conditions of their societies. Equally, their researches modified those conditions. Turning research in a particular direction introduced new techniques and new technologies, transforming the environments for later inquirers. Fashioning the conceptual tools required for pursuing particular lines of investigation achieved similar things in ways that were less overt but sometimes equally consequential.

As citizens and as scientists, people are heirs of decisions that might have gone differently. It is wrong to believe that, independent of any social history, scientific research would have delivered to us exactly the corpus of beliefs, concepts, problems, standards, apparatus, and technology we have today. A significant point from the sociology of science is that our science is contingent, and that if decisions about which human goals were worth pursuing had gone differently, both the science that we would have and the world in which we would live would be different.

Knowledge and Power

We want to close this chapter by suggesting that any philosophical account that takes the history and the sociology of the sciences seriously must address some important critical questions. The issues can be focused by thinking about a familiar slogan. Knowledge, we are often told, is power—and the telling is often intended to inspire gratitude for what the sciences, the most striking forms of knowledge, have done. As Francis Bacon envisaged at the dawn of modern science, the knowledge accumulated in physics, chemistry, biology, geology, and other fields of natural science has greatly extended our powers of intervention and control (although it is worth noting that the deliberate application of theoretical knowledge in technology only began to deliver serious fruits in the nineteenth century). Yet it is worth asking who "we" are, we who have been given this increased power, and also just what forms of power have been acquired.

Four important forms of power can be distinguished. First, and most obvious, is the power to intervene to reach goals people want (or to use predictions about the future to avoid consequences they fear). Just as evident is the fact that the powers of intervention are distributed very unequally across our species: The routine medical interventions that free the affluent world from many crippling and fatal diseases, to take just one example, are unavailable to many of the world's poor.

There are other types of power: power to control other people, power to enter into various types of discussions, and, most fundamentally, power to set one's own goals. When critics attack the sciences for providing "technologies of power," they often have one or more of these types of power in mind. They recognize the possibility of increased interventions in nature, but emphasize that the enhanced capacities are distributed very unequally. The skewed distribution makes it easier for some people to control the lives of others (even to do so without making the control recognizable by those subject to it), to exclude some voices and points of view from discussions that affect the character of the societies in which people live, and to narrow the range of options people see as open to them. To the idea that knowledge is liberating, the critics oppose the thought that it can also be used to confine.

Any philosophical account of the sciences needs to come to terms with the questions these critics raise. For the character of the knowledge we have is dependent on past social decisions. Many of those contingent decisions were made in contexts marked by just the forms of inequality we find in the contemporary world, where only some people reap the technological benefits the sciences have made possible, and only some have a serious say in how continued shaping of the social and natural worlds should go forward. Furthermore, it may be that past social decisions have introduced ways of limiting the ability of some groups and people to speak, to act, and even to think. Some of the limitations result when powerful people use scientific knowledge and technological tools to control or constrain what others do—some governments, for example, employ the knowledge and tools that science provides to control their citizens and stifle dissent; some corporations use similar means to control workers and stifle labor activism, and to shape the behavior of customers as well. The role that science can play in such contexts is a serious concern, and we consider it more fully in Chapter 6. Here we want to focus on the worry that the past decisions represented by science can constrain not just people's actions, but their thinking.

Consider an obvious example: the foundation, on facts about differences among members of our species, of a racist anthropology that brought serious consequences for those marked as belonging to inferior races—in particular, for their capacity to exert various forms of power. No doubt the elaboration of that anthropology involved all sorts of mistakes, misperceptions, errors of reasoning, and the downright falsification of alleged data, but it also proceeded by fashioning a particular conception of which differences among people matter, a conception that so permeated the world in which human beings lived that some of them were simply unable to perceive things to which they might have aspired as possible options for themselves. Those labeled as belonging to "inferior races" have been bound by scientific research far more than they have been liberated by it; the same may be true, in less obvious ways, of others as well.

The power to intervene in nature often really does enhance our lives, so long as it is not accompanied by any diminution in our capacity to decide what we want to do and what we aspire to be. If, however, the development of the sciences limits your vision of yourself, and what you might become, so that goals you might have pursued—and even seen as central to your life—no longer figure on the horizon of your decision making, then the increased powers of intervention are purchased at too high a cost. Science has given you refined abilities to obtain things you would have regarded as unimportant, if it had not simultaneously deprived you of the power to recognize what really matters.

To be clear, that is a critical accusation. It is neither unintelligible nor subject to easy refutation. Recognizing the social embedding of scientific research should bring home the ways in which past decisions pervade the present. Without examining those past decisions, we cannot tell how, and to what extent, they have expanded our horizons or limited our self-conceptions. So there is a serious empirical project of examining the intertwining of science, society, and self-definition—what some have called "tracing a genealogy." Undertaking that project might reveal that optimism about the liberating power of knowledge was entirely justified. A more mixed picture might indicate ways of improving the world that science—socially embedded science—has so profoundly shaped.

Suggestions for Further Reading

A pioneering study of the distinctive contribution of women scientists focusing on Barbara McClintock is Evelyn Fox Keller, *A Feeling for the Organism* (San Francisco: Freeman, 1983). Nathaniel C. Comfort, *The Tangled Field: Barbara*

McClintock's Search for the Patterns of Genetic Control (Cambridge, MA: Harvard University Press, 2001) is a biography of McClintock offering a different interpretation of her work. Keller's *Reflections on Gender and Science* (New Haven, CT: Yale, 1995) is a classic exploration of gender issues concerning scientific practice. *Feminism and Science* (edited by Keller and Helen Longino) is a valuable collection of articles; another is Janet A. Kourany (Ed.), *The Gender of Science* (Upper Saddle River, NJ: Prentice Hall, 2001). Donna Haraway, *Simians, Cyborgs and Women* (New York: Routledge, 1991) presents a radical vision of feminist science.

The transformation of primatology through the work of female primatologists is thoroughly described and analyzed in Haraway's *Primate Visions* (New York: Routledge, 1989). A highly readable first-person account is provided by Shirley Strum, *Almost Human* (Chicago: University of Chicago Press, 2001).

For a history of the Green Revolution research program in a larger policy context, see John H. Perkins, *Geopolitics and the Green Revolution: Wheat, Genes, and the Cold War* (New York: Oxford University Press, 1997). Vandana Shiva, *The Violence of the Green Revolution: Third World Agriculture, Ecology, and Politics* (London: Zed Books, 1991) gives an influential critique of its impacts. An illuminating overview of the global history of intellectual property rights in genetic material is Jack R. Kloppenburg, Jr., *First the Seed: The Political Economy of Plant Biotechnology*, 2nd ed. (Madison: University of Wisconsin Press, 2004).

A ground-breaking early exploration of the interactions between changing conceptions of gender, nature, power, and meaning in the Scientific Revolution can be found in Carolyn Merchant, *The Death of Nature* (San Francisco: Harper and Row, 1980). The roots of a Romantic scientific response to the "disenchantment" of the world, and some of its later influences, are traced in Robert J. Richards, *The Romantic Conception of Life* (Chicago: University of Chicago Press, 2002).

For models of scientific communities that reveal how apparently non-epistemic factors can produce a superior distribution of effort within a scientific community, see Philip Kitcher, "The Division of Cognitive Labor," *Journal of Philosophy*, 87, 1990, 5–22. David Hull's *Science as a Process* (Chicago: University of Chicago Press, 1988) explores the role of credit in the dynamics of science. Steven Shapin and Simon Schaffer offer a seminal rethinking of the Boyle–Hobbes debate in *Leviathan and the Air-Pump* (Princeton, NJ: Princeton University Press, 1985). Ian Hacking, *The Social Construction of What?* (Cambridge, MA: Harvard University Press, 1999) provides incisive analysis of debates about social constructivism.

A classic study of knowledge and power is Michel Foucault, *Discipline and Punish* (New York: Vintage, 1977); see also many of the essays in Foucault's *Power/Knowledge* (New York: Pantheon, 1980).

CHAPTER 6

Science, Values, and Politics

The Aims of the Sciences

For nearly a century, philosophical discussions of the sciences have concentrated on epistemological and metaphysical questions—When does a body of evidence support a hypothesis? Does microphysics provide the complete account of the cosmos?—to the neglect of issues about values. Yet, as Chapter 4 revealed, close attention to the history of science makes it impossible to view science as a value-free zone. Chapter 5 explored a number of criticisms of scientific practices, and many of the objections turn on considerations about what sorts of inquiry are valuable. Our present (and final) aim is to bring the discussion of values firmly and fully into the philosophy of science, so that important contemporary controversies can be faced, rather than being left to lurk uneasily on the philosophical sidelines. We begin with a straightforward question: What is the aim of science—or, perhaps, what are the aims of the sciences?

There are two traditional answers. One, explicitly formulated by Claude Bernard in the nineteenth century, was reviewed in Chapter 2—scientific research aims at explanation, prediction, and control. Scientific advances are pursued in the expectation that they will deliver increased understanding of nature, increased ability to predict events, and increased powers of intervening to produce desired outcomes. The second, equally common in discussions about science, proposes that science aims at truth. Let us start with this second response.

What would it mean to say that our goal is truth? Truth and falsehood are attributes of statements. Presumably, then, the idea is that science aims to tell us true statements. Which ones? An ambitious answer would declare that the intended end is the complete truth about nature—the truth, the whole truth, and nothing but the truth. That, however, is surely unattainable. No individual human being, or imaginable society of human beings, could write down the complete truth about our world. Nor would they want it—for an unordered list of true claims would not be worth having.

Perhaps what is wanted is an organization of the truth about nature, a complete account that assigns each particular truth a place (or does so "in principle") in an overarching "theory of everything." That would be something worth striving for, but the prospects are not promising (see Chapter 3). Moreover, there is a fundamental difficulty. In which language will the "truth about nature" be written?

You are reading this page in a local environment about which a vast number of statements are true. However speedy you are, time is continuous, and at each temporal point, there are infinitely many spatial relations to be determined, infinitely many other physical properties and relations to be sought. You might choose any of an infinite number of languages (an infinity at least as great as the power of the continuum) for expressing the truths about your environment, drawing the boundaries of objects and dividing them into types in any of infinitely many possible ways, and in each of these languages you could formulate an infinite number of true statements.

Most truths do not matter to you, or to anybody else. Unless you are planning to hang wallpaper, the exact variations in height around the walls of your room are of very little concern (and, even if paper-hanging is your métier, there are limits to the precision you will demand). Almost certainly, you will not aspire to know the exact number of blades of grass on the lawn (or to worry about how the exact boundaries of a single blade are to be drawn). Behind the thought that science is the search for truth is a presupposition that some truths matter. There are things that you—that "we"—want to know. Call these the *significant truths*. This yields a better version of the basic idea: Science aims at significant truth (or at true answers to significant questions).

Of course, this only raises a new issue: What makes a truth or a question significant? Here the first approach to the aims of science proves helpful. Significant truths are those that enhance "our" understanding or that enable "us" to predict events or to intervene in nature in ways that "we" want to. If, for the moment, we set on one side the goal of explanation, it is easy to envisage a pragmatic approach to the aims of the sciences, one that connects easily to views that have appeared in earlier chapters (Chapters 2 and 4). "We" have goals of predicting certain kinds of things, and of reshaping our environments in particular ways: The significant statements are those that enable "us" to achieve these goals; sometimes we require the exact truth, but frequently an approximation will do.

Our (irritating) use of quotation marks here is intended as a constant reminder that there are serious issues about understanding the words enclosed. Individual people and groups of people have goals, but not all individuals or groups have the same goals. Whose goals are to be given priority?

That question could be avoided if you envisaged, as some philosophers seem to do, that nature sets an agenda for scientific research. Yet, when we take it seriously, the idea that particular phenomena are privileged for exploration—as if they shouted out "Predict me!" or "Intervene in me!"—looks deeply problematic. Prediction and intervention are activities that must be focused in particular areas of nature and directed toward particular goals, and the only ways in which direction can be given stem from the aspirations of human beings. Reference to people and their aims is essential. So we cannot avoid the question "Which people?"

The actual course of scientific research, throughout history, has surely reflected the goals favored by select groups of people: individual investigators, communities of investigators, political and military leaders who have recognized the possibilities of power-yielding knowledge, and entrepreneurs with similar appreciations. Whether the actual course of history corresponds to the way things should be is another matter. There is a normative issue: Whose goals *ought* to be considered in specifying which predictions and interventions give rise to significant questions?

So far, we have proceeded on the assumption that only pragmatic issues matter—we explicitly set on one side the potential aim of explaining nature. Perhaps this was too pragmatic. Many, probably most, scientists would vehemently protest our simplification, contending that although increased abilities to predict and to intervene might be useful spinoffs from scientific advances, the central achievements lie in the provision of greater understanding of natural phenomena. Without supposing that this understanding is reached by working out parts of some ideally unified picture of the world—pieces of the eventual "theory of everything"—it is possible to suppose that, just as certain kinds of predictions and interventions matter to "us," so also do particular questions, questions that provoke "our" curiosity whether or not their answers have immediate applications to practical projects. Finding out what happened in the first milliseconds of the universe or discovering the exact relationships among various hominid species may have no bearing on human capacities for prediction and intervention. Nevertheless, "we" may be intrigued by past events, and be glad that the sciences inform us about them.

One position on the aims of the sciences, perhaps one that many scientists would defend, proposes that the central questions of different fields arise out of disinterested human curiosity. These questions are significant because people naturally wonder at some natural phenomena, at the regularities of motion of the heavenly bodies or the uniform ways in which seeds grow into particular types of plants. Technical problems in the sciences arise from efforts to answer these broad questions, and the results should eventually deliver to all thoughtful people explanations that satisfy their curiosity. Because scientists are in the best position to understand how the phenomena that excite "our" wonder are best pursued, they are the best judges of what counts as significant.

Is that correct? Are there intellectual goals of understanding that all people should value? Even if there are, do the values ascribed to these goals override all other concerns that people outside the scientific community might have? Should those people be content to let that community decide what research should be done, how it should be pursued, on what basis conclusions should be accepted, and how those conclusions should be applied? In a world in which scientific innovations often have commercial possibilities, research is increasingly shaped by business managers—is that a good thing or something about which outsiders (as well as scientists) should worry? Value judgments seem to arise not just as expressions of a hypothetical curiosity shared by all, a panhuman sense of wonder, but also in the thought that satisfying this curiosity is a paramount task. They also enter into the further decisions about how research is properly to be done, how conclusions are to be drawn, and how results should be translated into practical action. Is the scientific community uniquely qualified to make those value judgments on behalf of the rest of our species (indeed, on behalf of our planet, and all its inhabitants)?

Values and Choices

The various critiques of science considered in the previous chapter provide a useful starting point for thinking about the value judgments science requires, and about who should be making them. Science (the critics point out) in fact treats certain kinds of truths as important or interesting; it is responsive to the needs and goals of certain kinds of people; it is animated by certain kinds of values (particular values that are contingent features of the way scientific inquiry is carried out). If nature itself picked out the significant truths, we might suppose that they would be equally useful to all people, and pursuing them would be equally compatible with all values, but nature is incapable of telling us where to investigate. Choices have to be made—closely interconnected choices about what we want science to do for us, which truths matter, and whose voices are heard. Critics of scientific practice play the useful role of urging attention to the choices that are actually made, and asking if they are defensible.

One striking feature of the choices made in both historical and contemporary science is what they leave out. The various critiques of science agree in pointing to two broad categories of truths that the sciences have tended either to overlook or actively to exclude from consideration: truths about qualities, and truths about wholes. These omissions are not unmotivated—on the contrary, they stem from two ideas that were crucial to the groundbreaking early successes of the natural sciences: the idea that science should treat only those properties that could be quantified or measured, such as size, mass, and velocity; and the idea that science should proceed by means of analysis, seeking to understand wholes by looking at their parts. The focus on quantitative properties was intended to ensure that scientific knowledge would be objective and universal: objective because measurable properties like size and velocity were thought genuinely to belong to physical objects and their ultimate parts; universal because measurement can reveal mathematical patterns in the phenomena that constitute the "laws of nature." The exclusion of the unquantifiable from science means, however, that certain kinds of knowledge become invisible, including the kind of fine-grained knowledge about the concrete qualities of local ecologies and geographies that is characteristic of traditional societies. For some people, this kind of knowledge may seem most significant, encapsulating the details of locally important things and processes.

The reductive strategy of analysis is another means of achieving universality, as it can reveal the common components that make up apparently diverse kinds of objects. It has proved to be a powerful way of learning how to intervene effectively in the world. The rapid advance of technology, from the mechanized pumps and looms of the Industrial Revolution to today's sophisticated information technology and biotechnology, was made possible by systematic efforts to break complex causal systems down into component causes that could be moved from one setting to another. Yet overemphasizing the reductive strategy has also fostered neglect of complex causal interconnections at work in functional systems like organisms, ecosystems, and economies (recall the Balinese water temples in Chapter 4).

Why is the reductive strategy so attractive? Partly because of the dominance of a particular value. Some scientists, and many of the people making decisions for the governments and businesses who pay for most scientific research, have been motivated by the promise of new capacities for intervening in the world—the promise of control. They value universality, which ensures that the capacity for control will be exportable from the research site to the sites of many different applications. They also value knowledge that has a market worth—proprietary knowledge that helps to produce patentable devices, materials, organisms, or processes that reliably produce desirable effects. Critics point out that other values could guide research along quite different paths from those it now follows. Many small-scale efforts are being made to produce knowledge that is intended principally to support stable communities and sustainable economies, foster cultural continuity in the face of new discoveries, contribute to an intellectual commons, or help disadvantaged individuals or communities to identify and realize their own goals. These projects make up only a tiny fraction of the research that affluent societies now choose to pursue. What would a science seriously focused on serving such excluded values look like? What could it achieve? From our current position it is very difficult to tell. Would it be worth experimenting, to find out?

Science excludes or marginalizes certain truths and certain values, in part at least, because it excludes or marginalizes certain kinds of people. The institutional structures of science keep many people from being able to participate fully in making scientific knowledge, or in deciding what science should be done. Women, poor people, and people who are culturally and physically remote from the centers of scientific activity and policymaking are underrepresented among scientists and among those who make decisions about science. Their perspectives, needs, and interests are often neglected; indeed, both scientific knowledge and science-based technology have too often been used in ways that are actively detrimental to their interests.

The multifarious specific claims made by critics of science will continue to be debated for many years, no doubt, but there is a common core to their work that demands serious consideration in the meantime. The root of the critiques of Chapter 5 is an attack on the pretense of value freedom. Once the faulty—but influential—philosophical conceptions of science as value-free and context-independent have been challenged, the way is open to consider, and perhaps to contest, the value judgments underlying the choices made in the course of scientific inquiry. Those value judgments are reflected in the questions commanding scientific attention and in the uses made of scientific findings. With respect to such choices, critics ask two questions, the urgency of which is undeniable: What is excluded? Who decides?

The Autonomy of the Sciences

Who should make decisions about the direction that scientific inquiry takes? There are many reasons to think scientific inquiry can and should be made more responsive to the goals and values of a broader public. But this idea runs counter to the long-standing tradition that upholds the *autonomy of the sciences*: the view that outside direction distorts or impedes scientific inquiry, that science thrives only when scientists are free to "follow the truth" (or the path they deem most likely to lead to the truth) wherever it leads. As it stands, this version of the idea of scientific autonomy inherits the flaws of the value-free ideal. Which truths are worth following depends on what goals are in view, and there are good reasons to doubt that scientists are in a position to make authoritative judgments about the goals toward which scientific research ought to be directed.

A more modest form of autonomy might still be defended, however. If there were a more inclusive way of determining the goals that inquiry is to serve, one that involved a broader public in the decision-making process, scientists might then be in the best position to compare the prospects of possible research paths to those goals and judge which approaches are best suited to achieving them. Even this more modest kind of autonomy becomes less plausible, though, when we consider how incomplete the perspective of any individual scientist is. Suppose that one goal is an increase in agricultural production. Geneticists, chemists, and ecologists will have very different recommendations about what kind of research to pursue. How should the disagreement be adjudicated or their various proposals integrated? Furthermore, the shared goal—increases in agricultural production—is itself vague, requiring specification and adjustment as the process of inquiry goes forward. But this means that the broader public needs to be involved in the goal setting not only at the beginning, when some ill-defined target is defined, but along the way, as possibilities for specifying it in rival ways are being discussed.

Is there nonetheless a grain of truth in the ideal of autonomy? Is some kind of autonomy possible, and important, in scientific inquiry despite the obvious challenges it faces? To answer these questions, we need to think more concretely about the relationships between scientists and the larger society. The argument for autonomy is often made by pointing to a few well-worn (and extreme) examples of the fruitfulness of free inquiry, and of the harm done by external direction. It is worth looking a bit more closely at the range of real possibilities, however, and especially at the actual state of affairs.

Powers Behind the Lab

Those who hold political power have often found it worthwhile to sponsor scientists' work, and have often sought (naturally enough) to direct and constrain its courses to suit their own ends. This has been true since science's earliest days, but the stakes changed dramatically in the twentieth century for both scientists and their sponsors. The rapid advance of science-based technologies crucial to success in warfare, agriculture, industry, and medicine, together with the increasing relevance of scientific knowledge to issues of public policy, gave governments powerful new incentives for seeking control of scientific inquiry. At the same time, the rise of professional science, and especially of "big science"—scientific research programs requiring such massive investments in equipment and personnel that only national governments could support them—gave governments new kinds of leverage in their dealings with scientists. Several distinguishable kinds of relationships between governments and scientists have emerged as a result.

One important kind of relationship seems at first to follow the lines of the more modest form of autonomy sketched earlier. Governments specify goals in the form of certain products—objects, processes, capacities, or information—that they want to acquire, and the task of scientists is to produce the knowledge needed to create those products. When the governments in question are representative democracies, this might appear to be an excellent division of labor: The voices of the citizens are given their proper place. Yet it is evident at the outset that, at best, only the citizens of a particular state are likely to be represented, and that—given the difficulties citizens face in comprehending technical scientific issues—even within this limited domain, representation will probably be imperfect. Further, scientists often take the lead in the decision making: They are the first to see the practical possibilities hidden in their theoretical work, and propose applications worth pursuing; governments then choose which of these will be funded. A great deal of "big science" in the last hundred years has been organized on this sort of basis: Weapons research, the space race, and many research programs in energy research and communications technology offer some obvious examples.

Advocates for the scientific community have famously—and influentially—argued that this kind of relationship, in which research in applied science is done "to order," can succeed only against a background of basic research directed by scientists in a fully autonomous way. (The most well-known articulation of this conception of the relationship between the instrumental goals of government and the epistemic goals of scientists was made by Vannevar Bush in 1945; it helped shape the National Science Foundation in the United States and indirectly influenced the structure of many national funding agencies elsewhere in the world.) These relationships are celebrated for approximating the ideal of autonomy where basic research is concerned. But what do they really accomplish? Notice that where resources essential for research are at stake, scientists have strong incentives to align their goals, and the proposals they put forward, with the interests of the governments that they depend on for the desired "no-strings" funding of basic research. Their judgments about what questions are most important may be subtly or not-so-subtly distorted as a result. The scientists' autonomy falters, not because it is overridden by governmental fiat, but as a result of normal features of patron–client relationships. Or, perhaps, it becomes a deliberately deceptive venture in which promises are made that just the sort of research the grant-writer wants to pursue will deliver—quickly—exactly the results the government sponsors want. So, for example, the plan to map and sequence the human genome was often advertised as promising speedy relief to people who suffer from genetic diseases. The real benefits of the research—considerable for advances both in basic research and diagnostic medicine—were seen as an inadequate basis on which to request government funding. Is the autonomy of the scientific community preserved when scientists are forced to disguise the work they want to do or to inflate the practical benefits that are likely to flow from it? (See "Payoffs from Genetic Sequencing".)

∽ Payoffs from Genetic Sequencing

When scientists sought funding for the Human Genome Project (HGP), they emphasized the benefits for health. The most ambitious suggestions contended that, once the techniques of sequencing were refined so that "reading" anyone's genome became a manageable—even a commonplace—task, it would be possible to cure or treat many diseases that resist intervention. Critics warned that promises of instant cures were overblown, and many analysts who championed the HGP agreed that the real benefits lay elsewhere.

Given the rapid progress of sequencing technology, fueled by work on the HGP, nobody should regret the investment in it. Explorations of non-human genomes have disclosed many surprises, especially in revealing the complex ways in which genetic regulation works. Genomes are now understood as dynamic systems, in which a given sequence can be transcribed in different ways, depending on background conditions in the cell. Comparisons of the genomes of different species have not only illuminated their evolutionary relationships, but also opened up fruitful avenues for exploring questions about development and cellular metabolism. The HGP has turned out to be extremely valuable for basic research in biology—as the scientific enthusiasts were confident in advance that it would be. The knowledge and pointers for research it has already generated will surely contribute to medical advances in the future, even though we cannot confidently predict what they will be or when they will come.

Yet the critics (and the analysts who agreed with them) were correct. Immediate payoffs for the cure or treatment of disease are rare. If you understand that a particular sequence at a particular locus predisposes the bearer to a particular disease, that doesn't necessarily mean you can do anything to prevent, or even mitigate, the problem. Perhaps knowing the sequence will give you some insights into some step in the process that leads to the malfunction of an organ—you might see, for example, that the molecule required to do a specific job might be missing. Perhaps if you got that far, you could do something to remedy the situation—maybe by injecting an appropriate substance at the right stage. Unfortunately, these possibilities turn out to be very uncommon. Sequencing typically provides increased ability to diagnose, but rarely translates quickly and directly into successful treatment.

Speed and precision in diagnosis are often really beneficial. Cystic fibrosis (CF) was one of the first diseases to be understood from the genomic point of view (in this case, the breakthrough came before the HGP). CF afflicts about one person in 2,500 (among people of northern European ancestry; the frequency is less among people of different ethnicities), and used to cause death in childhood or adolescence. Thanks to the ability to identify CF children early, their care can be managed much more effectively, and the average

(cont).

life span for CF patients is now in the upper thirties. The increase is partly traceable to improved treatment (itself inspired by the molecular understanding of the disease—this is one of the rare successes), but, even in the absence of that treatment, prompt diagnosis allows the crises of early childhood to be overcome without severely damaging aftereffects.

In many instances, however, diagnosis only provides doctor and patient with advance warning about what is coming. Molecular understanding of neurodegenerative diseases—like Alzheimer's and Huntington's diseases—allows people who suspect that they are at risk to be tested, but, once they know more about the probability that they will be afflicted, there is nothing they can do to improve the odds. Some go ahead with the test, believing that the knowledge will be useful in planning their lives. Many do not, because they do not want to live under the shadow of the future.

Sequencing has other potential social uses—in the detection and prosecution of crime, for example—but its principal benefits have widely been seen as medical. Those benefits may eventually come, and, a century hence, historians of medicine may praise the wisdom of investing in the HGP, recognizing it as beginning a development of basic biological knowledge that transformed medicine (as historians today see the early genetic researches of the twentieth century as the foundations of the molecular medicine we now have). Yet it was always predictable that the payoffs to patients would come at a slow and uncertain rate. After all, the molecular basis of sickle-cell disease (a condition that disproportionately afflicts people with recent ancestors from Africa) was identified forty years before the HGP was conceived—and effective treatment for it is still not available.

Plainly, there can also be more obvious invasions of scientific autonomy. Sometimes, scientists investigate matters with important implications for policy or for points of ideological or religious doctrine. Here, political authorities favoring particular policy choices or ideological positions can put pressure on scientists to draw (or at least to publish) only conclusions consonant with the views or outcomes that they prefer. Some cases of pernicious political interference in science are legendary. In 1616, Galileo was forbidden to continue to publish his arguments for a Copernican cosmology. Within a decade, however, one of his friends had ascended to the Papacy, and Galileo was allowed to articulate his views in a book, *Dialogue Concerning the Two Great World Systems*, which was submitted to the office of the Papal Censor in 1631. After the text had been scrutinized, Galileo was summoned to Rome, shown the instruments of torture, and forced to recant—but it was too late, for copies had already been smuggled out of Italy, and had begun to do their subversive work. Even darker is the case of Stalin's support for the disastrous enshrinement of Trofim Lysenko's theory of inheritance. Not only did this lead to the execution or imprisonment of eminent Russian scientists who had defended Mendelian genetics (which was denounced as a "bourgeois pseudo-science"); it helped bring about widespread famines caused by the recurrent failure of Soviet agronomy (see "Lysenko and 'Planned Science'"). To these two celebrated instances, we can now add the far milder—but perhaps even more consequential—interference, carried out by the administration of George W. Bush, in the reporting of climate science.

ᴄᴏ Lysenko and "Planned Science"

Defenders of autonomy in scientific research often cite the "Lysenko affair" as a cautionary tale about what happens when governmental planning dictates the course of scientific research. Trofim Denisovich Lysenko, a previously un-known agronomist, rose to prominence in the late 1920s, when his treatment of seeds promised to improve crop yields. After some years with poor harvests, authorities in the Soviet Union were concerned to increase the production of wheat, and they were impressed with Lysenko's experimental program. With-out any deep knowledge of the underlying biology—and, in particular, of the genetics that had been developed by Morgan in the wake of the rediscovery of Mendel—Lysenko proposed that it was only necessary to treat the first genera-tion of seeds, and that the effects of the treatment would be inherited, a pro-posal at odds with classical genetics, which had rejected the inheritance of acquired characteristics (Lamarckism).

During the 1930s, Lysenko rose to prominence, and his ideas and pro-grams became official Soviet policy, and even official Soviet science. Although the agricultural results were often disappointing, Lysenko was adept at offer-ing new "improvements," and he retained the support, not only of Stalin, but also of Stalin's successor, Khrushchev. The exposure of his errors came only in the 1960s. By then, a large number of agricultural plans had failed, leading to food shortages that surely caused many deaths (how many exactly is im-possible to determine). Moreover, Lysenko's resistance to orthodox genetics led to denunciations of prominent scientists: Not only was Soviet biological research profoundly damaged, but some brilliant biologists were prosecuted and imprisoned (some of them died in prison).

The Lysenko affair shows government planning of science in a very strong sense. Not only did the Soviet authorities decide what scientific prob-lem was to be addressed, but they specified what was to count as "scientific knowledge" in addressing it. Beyond that, they were ruthless in enforcing the *(cont)*.

orthodoxy on which they had (unwisely) decided. Interference of this type should be distinguished from "planned science" that restricts itself to identifying important problems for scientific research and to providing incentives for researchers to work on those problems. Governments have often engaged in planning in this weaker sense—as in World War II, when the United States instituted the Manhattan Project and the British government brought a team of investigators together to crack the German Enigma code. Indeed, government planning is a commonplace in recent and contemporary science, in the "war on cancer," the "decade of the brain," and the Human Genome Project—not to mention in the standard ways in which scientists compete to obtain public funding. There is a world of difference between supposing that the direction of scientific research should be shaped by outside needs and by democratic values, and the extremes of Soviet "Lysenkoism" and of the "racial science" of the Nazi doctors.

There is an important distinction between directing lines of research and dictating conclusions. The mandate of democratic governments is to represent the public's preferences and to secure the public good. It is easy to see that such governments have legitimate authority to make at least some decisions about the direction pursued by publicly funded science, although they are obliged to make such judgments in light of the best information available to them, which often must come from the scientists themselves. Even democratic governments cannot, however, legitimately tell scientists what to conclude. Having determined what questions the public wants and needs to have answered by science, governments are obliged to allow scientists in fact to answer those questions. Yet despite the importance of this contrast, the line may be hazy in practice: A narrow enough specification of what to investigate can amount to a specification of what to conclude. During the era of Lysenko, the permissible questions for further research were framed by his idiosyncratic (but ideologically congenial) assumptions. No easy rules can sharpen this line for us; it can be located and defended only by means of critical discussion. Who should be party to that discussion, and what standards it should employ, is part of the larger question of how, and by whom, the aims of the sciences should be chosen.

The meliorative project about science, pursued by commentators and critics who hope to improve scientific practice, calls for a reconsideration of the governance of science: of who determines the aims of the sciences, by what means, and in keeping with what standards. Governments have sometimes played an important role in ensuring that some science serves the good of relatively broad publics (consider the way in which the U.S. government responded—eventually—to the spread of AIDS); undoubtedly they will continue to have an important role to play. We can see reason to doubt, though, that government alone is a sufficiently sensitive tool for determining the aims of the sciences. Particular governments (and their individual officers) inevitably have interests that are distinct from the public interests they are charged with advancing. Moreover, the danger of bias resulting from the alignment of the interests between the makers of knowledge and the wielders of power is a serious one. Finally, even at their most responsive and most responsible, democratic governments represent only a fraction of the human population.

If governments cannot be relied on to ensure that science responds sensitively to the diverse and shifting needs of the world's people, perhaps the invisible hand of the market can do the job. Increasingly, science is undertaken by investigators employed by private industry, and this work is indeed directed more by the market than by any considerations of policy, although, of course, policy constrains it in certain ways. Markets can function as a democratizing mechanism, allowing the individual choices of buyers and sellers to contribute to a collective decision-making process. Can they work this way for science?

What does market-driven science look like? Above all, it must focus on making discoveries that have a market value, on knowledge that can be packaged as a salable (preferably as a patentable) product. Of course, this means that privately funded science is overwhelmingly "applied" rather than basic; much of it is even more narrowly focused on the details of product development. But the implications reach further than this. Because the entrepreneurs who support research want to achieve large economic returns, they impose mechanisms that disrupt the free flow of scientific information. In particular, patents can make certain lines of investigation effectively impossible: Contemporary ecologists are sometimes hampered because their serious questions about interactions in the environment require them to trace a long sequence of patents previously acquired by agribusiness concerns; because there are so many different potential patent holders (who hold the title to particular DNA sequences), they recognize that even the most thorough legal inquiry (which, of course, they are rarely in a good position to conduct) might leave them vulnerable to some future suit; so, reluctantly, they abandon their project. Moreover, market-driven research tends, inevitably, to concentrate its efforts in areas likely to produce outcomes (i.e., products) of interest to promising markets: to wealthy consumers, and well-funded public enterprises such as military organizations. When aimed at the consumer market, it tends to focus on those aspects of (well-off) people's lives that they are most willing to spend money on—however trivial or base these may be.

These consequences are bad enough. Perhaps even more worrisome, however, is the incentive markets create for scientists to pursue research aimed not principally at fulfilling people's needs or desires, but simply at maximizing entrepreneurs' profits—even at the cost of consumer well-being. Tobacco company scientists have been accused of seeking ways of making cigarettes even more addictive, but that is just what the market would be expected to ask of them. Even where market-driven research does address pressing public problems, it must systematically neglect lines of research that could lead to solutions that happen not to be profitable to the industries that fund the research, or that undermine their existing sources of profit. Privately funded medical researchers focus their efforts on proprietary pharmaceutical treatments—although these may be costly to the public and have undesirable side effects—and they may overlook equally effective "lifestyle" interventions. Privately funded geneticists have focused on developing crop varieties that perform well only in combination with the proprietary fertilizers and pesticides sold by the seed companies, although poor farmers would benefit from the development of improved varieties that do not require expensive inputs. As with government-sponsored science, there is a serious worry here that the bias resulting from these market-driven choices affects not just scientists' actions, but their beliefs and judgments; that industry scientists unconsciously align their own values—and so their evaluations of epistemic significance—with the interests of their employers. This would be a serious failure of autonomy. Defenders of the ideal of autonomy rarely discuss privately funded science, perhaps because they think of it as aiming simply at finding practical applications for the significant truths discovered by publicly funded basic science, or perhaps because it seems self-evident that private funders should be able to control the research they are buying. The cumulative effects of market-driven research need to be considered, however: What is accepted today forms the basis for all sorts of evaluations that will be made tomorrow—judgments about what lines of future research are valuable, and what sorts of conclusions are well-supported. People, knowledge, and values move across the private–public boundary.

Perhaps the most troubling form of inappropriate direction of science comes in the form of corruption, when scientists lie or deliberately distort their methods to produce particular results desired by their sponsors. Such corruption is continuous with the kinds of alignment of interests we noted earlier for both governmental and private funders. No doubt a continuum exists between relatively innocent bias and deliberate deception. Patterns in publication suggest that bias, at least, is commonplace—researchers investigating the safety or effectiveness of a product are far more likely to draw favorable conclusions if their work is funded by the company that makes the product—and deliberate deception has been uncovered often enough for concern. Of course, similar misbehavior can take place in the absence of any external interference, motivated by scientists' own desire for professional advancement or recognition, or even by their willingness to "improve" their data to convince others of a conclusion they believe to be true. But where essential funding or institutional standing is at stake, scientists are especially vulnerable to such temptations, whether or not there is any effort to influence them on the part of their sponsors.

Today's science is far from autonomous. If the ideal of autonomy is used as a justification for a laissez-faire approach to the governance of science, what we will get is a continuation of the form of nonautonomous direction that now exists: ill-considered, often unrecognized, and sometimes seriously harmful to the epistemic integrity of the sciences. If we presume that private science is legitimately directed by its buyers, but public science should be autonomous, the result will be a science strongly skewed toward the private interests of entrepreneurs and their wealthier customers. If we accept instead that autonomy is not an option, then we can do what needs to be done: to think seriously about who should be making decisions about the conduct of science, and on what basis.

What Do "We" Want to Know?

One way of approaching the notion of significance that underlies any viable account of the aims of the sciences is to take the references to "us" inclusively: Us means all of us, all people, past, present, and future. The goals are determined by what this vast community wants to know, perhaps because all members of it share some sense of wonder at particular aspects of nature, perhaps because there are challenges that demand specific types of prediction or intervention. A democratic approach to the aims of the sciences would insist that the goals are determined by the wants of all.

Vulgar democracy supposes that all of us have our personal desires for knowledge, and, where there is disagreement, there is nothing for it except to vote. There are good reasons for thinking that vulgar democracy would be a disaster for the sciences, that it would create a tyranny of ignorance that would curtail or eliminate lines of research that might eventually satisfy a broad spectrum of human needs. Would the public have thought it a good idea in 1910 to explore problems of heredity by breeding fruit flies? At that time, some prominent theorists of heredity campaigned for direct application of the recently discovered Mendelian ideas to human genetics. They supposed the new insights would swiftly give rise to methods of treating hereditary diseases. By contrast, T. H. Morgan and his students (later colleagues) pursued a more indirect path, concentrating on fathoming the genetics of a tractable organism (the fruit fly, *Drosophila melanogaster*), in hopes that this "basic research" would eventually yield tools for improving human health. Decades later, the promise of Morgan's work was fulfilled in contemporary biomedicine. It is easy to understand how myopic commentators, earnestly seeking the public good, might have dismissed Morgan's program at the very beginning.

To appreciate the troubles of vulgar democracy does not mean adopting the elitist view that the scientists know best how to identify what sort of knowledge is valuable. Democracy can be made more sophisticated. We suggest an ideal of *well-ordered science*. Scientific research would be well ordered if (and only if) it met a number of conditions. These conditions involve the concept of an *ideal deliberation,* a concept that must be explained before the conditions can be stated.

An ideal deliberation is a discussion among representatives of the different predicaments and perspectives found in the inclusive human population (i.e., our entire species, past, present, and future). Those representatives are required to readjust the wishes with which they come to the discussion, by taking account of the best available information about nature and about the prospects for research of different kinds, and by recognizing the equal worth of their fellow discussants and of their perspectives and preferences. Informed by what is known, and perhaps by what lessons can be gleaned from the history of inquiry, they are also mutually engaged with one another; that is, they aim to realize—insofar as it is possible—the aspirations of each of the others. They endeavor to reach consensus on how research should be directed (what it should aim at), how it should be conducted, what standards should be used in adopting potential new items of knowledge, and what uses might be made of the knowledge that research delivers. If, even under these ideal conditions, consensus cannot be achieved, the deliberation is forced to close either by adopting some conclusion that is at least acceptable to all, or, if all else fails, with a majority decision.

Given this concept of an ideal deliberation, the conditions on well-ordered science are as follows:

1. The lines of inquiry to be pursued are those that the ideal deliberation would endorse.
2. The modes of pursuit of those investigations accord with standards that the ideal deliberation would accept.
3. The judgments about which findings to incorporate within the evolving store of accepted scientific claims accord with standards that the ideal deliberation would accept.
4. The applications to be made of scientific knowledge would be endorsed by the ideal deliberation.

The ideal of well-ordered science introduces democratic values into the practice of scientific investigation, without submitting research to a tyranny of ignorance. It is, we suggest, an ideal it might be worth trying to realize.

Even if the ideal appeals, we plainly cannot try to achieve it by instituting a real-world counterpart of ideal deliberation. Public involvement in the four kinds of decisions to which our conditions respond might be more varied and more extensive than it is, but it would always fall vastly short of a forum in which all were informed, all mutually engaged, and all heard. Yet the limited democratization of scientific practice that we might, in practice, be able to achieve is often suspected as likely to do more harm than good. *Ideals* are all very well, but, as a practical matter, it might seem best to be elitist and let the experts decide.

There are obvious grounds for worry about democracy, even in its non-vulgar form. Basic research will be slighted, it might be thought, or scientific creativity stifled. Research is essentially unpredictable. On one or all of these bases, many scientists are inclined to reject any steps toward well-ordered science.

Yet there are straightforward answers. If basic research is significant, as it often is, then it ought to be possible to show both how and why it is significant. We pointed to one source of evidence earlier, in noting the long path that led from Morgan's indirect approach to medical genetics to the successes of today. The history of science and its applications provides us with a laboratory from which serious scholarship could develop a valuable picture of the ways in which the search for "pure understanding" prepares for later practical successes. Likewise, if letting "proven talent" go its own way is a valuable strategy for scientific investigation, then that, too, is something that could be shown. Detailed study of the history of science would reveal to us just how useful it is to let people with outstanding track records follow their hunches, rather than trying to direct research toward concrete problems. Nobody who favors the ideal of well-ordered science should object to taking into account what we can learn from past ventures in shaping the direction of inquiry. What defenders of that ideal should question are casual assumptions backed by no serious statistical study, but rather by the sort of evidence—familiar anecdotes—that would be ridiculed in professional scientific discussions.

The final complaint makes a serious point. With respect to human affairs generally, it is frequently difficult to be confident in our powers to predict. Any proposal for conducting research, whether committed to the ideal of well-ordered science or to something else, must acknowledge the fact that the trajectory of inquiry is very hard to foresee. Yet, once we understand that this is of a piece with many other human phenomena, we recognize that unpredictability does not issue a license to throw up our hands in despair, to toss coins, or proceed by whatever means tradition happens to favor. However unpredictable future events may be, when they are consequential we do—or should do—the best we can. People assemble what information is available, try to find patterns for prediction, and then go forward on that basis. (Think about how you make plans for your own future and for that of your family.)

Thoughts about the democratization of science can be focused by considering places at which contemporary practice appears to diverge from what would occur under well-ordered science. Consider biomedical research. Most of this is done in affluent countries, whose citizens are not subject to many types of infectious disease. Consequently, there are many diseases that afflict the world's poor, for which no remedies that might be exported to the environments of the sufferers are currently available, and virtually no current research is aimed at finding such remedies (see "Disease Research and Global Health"). There are many forms of respiratory infection, intestinal disorders, and parasitic infestations that now kill or disable massive numbers of people throughout Africa, Asia, and South America. The overwhelming majority of these diseases are strikingly understudied, even though some of the most egregious cases have recently been addressed by the welcome efforts of charitable foundations. If you suppose that, subject to roughly equal promise of progress, medical inquiries should be directed in proportion to the burden of death and disability that diseases cause, the current research agenda remains extraordinarily skewed. This is one clear departure from well-ordered science.

ᖆ Disease Research and Global Health

Among the poorer nations, the major causes of death are infectious diseases: respiratory diseases, forms of diarrhea, HIV/AIDS, malaria, and tuberculosis. Each year, millions of people, many of them children, die from these diseases. In 2008, the last year for which the World Health Organization has compiled statistics, almost 9 million people died from "neglected diseases" (the standard term for diseases that do not attract significant research funding). These diseases also bring a heavy burden of disability, and contribute to the difficulty of escaping from the cycle of poverty.

Disease research is supported largely by private corporations (pharmaceutical companies) and by government agencies (like the National Institutes of Health in the United States). Corporations primarily direct their research toward products that can be brought to market and sold for profit: It should not cause much surprise that they are not primarily concerned with the diseases of the world's poor, or that they see more future in diet pills and cosmetic creams than in treatments for Chagas disease or river blindness. In the affluent countries where most medical research is carried out, the budgets assigned to tropical infectious diseases are typically very small. A few private foundations (e.g., the Gates Foundation) have begun to recognize the problem of neglected diseases, and, as a consequence of their dedicated efforts, the research situation has improved somewhat in the past few years.

(cont).

The burden of a disease can be measured in various ways. The simplest and crudest measure is to focus on the percentage of deaths the disease causes, but that measure fails to take into account the suffering resulting from conditions that are not fatal. A more sophisticated measure is the disability-adjusted life year (DALY), which takes into account both the years of life lost because of premature death and the years lost through living in states of compromised health. The burden of a specific disease can then be calculated as the percentage of DALYs due to the disease (the ratio of the total number of DALYs due to the disease to the total number of DALYs due to all diseases, multiplied by 100). (Even this measure does not take account of the indirect harm a disease causes to people who do not themselves have it, such as children orphaned by AIDS.)

One way to articulate the idea of well-ordered science in the context of disease research is to suppose that well-ordered science would endorse the Fair Share principle: If two diseases are estimated as roughly equally promising as targets of research, the funding devoted to them should be proportional to their disease burden. According to the Fair Share principle, if the burden of acne is one thousandth of the burden of river blindness, then research on river blindness should receive 1,000 times the support that research on acne does (assuming, of course, that both problems are roughly equally tractable). We do not know any reliable estimates for the ratio of the burdens of these two diseases. We strongly suspect, however, that the Fair Share principle is not satisfied in this particular case.

A second departure from well-ordered science can be identified if we consider again a popular thought about the goals of the sciences. Many practicing scientists believe that their research is important because it will ultimately contribute to the satisfaction of human curiosity. If there were any approximation of an ideal deliberation, then we might be able to find out whether the hypothetical shared sense of wonder is real, and whether people consider responding to it more important than pursuing other lines of inquiry. Yet, even if we were to be convinced that some forms of curiosity are widespread among the world's people, and that scientists have identified them, it is abundantly evident that, in the present practice of science, far too little is done to translate the knowledge of the experts into knowledge for all. If the great benefit of science is that it increases "human understanding," then all of us should have access to that understanding. Otherwise, the sciences will look suspiciously like esoteric forms of art, developed only for the cognoscenti. Perhaps, if they were asked, a majority of members of our species would agree that refining the understanding of a small minority was a worthy goal, but that question has not been put to them, and, until it has, it is reasonable to suppose that the value of "enhanced understanding" presupposes that the enhancements are widely available.

Perhaps by reflecting on instances like these, where there is a discernible gap between the ideal of well-ordered science and actual practice, philosophers will be able to see how to take incremental steps in the direction of the ideal. Promising first steps have been taken in some special contexts. When decisions must be made concerning research on issues of great importance for a particular local community or regional population, it is possible to design a process of deliberation that brings scientists and different local people who have a stake in the outcome together to decide which questions need to be answered, and by what means. Such exercises have been found particularly useful for guiding the research that will inform decision making in conflict-ridden areas of ecological management and in other areas where different local groups differ widely from each other and from scientists both in their values and in their beliefs. Interestingly, well-designed deliberative processes in situations like these can help resolve some problems we have noted in other chapters. They can help bring the different value judgments of scientists and local stakeholders into better alignment, enabling them to identify common goals for inquiry, and foster the growth of trust that allows people to learn from each other. They can thus provide opportunities for scientists to learn from nonscientific experts who may have nuanced local knowledge that the scientists lack, and for local people to learn about the broader theoretical perspective that the scientists can bring to bear.

These very local forms of deliberation could provide models that can help us to design processes and institutions for democratic decision making in larger arenas. Methods of public deliberation that have recently been developed for use in broader contexts of political decision making, such as deliberative polling and citizens' juries, also offer promising models of ways in which representative groups of citizens, having been led "behind the scenes" to learn about the scientific situation through substantial dialogue with scientists, might work together to reach well-considered decisions about the direction of further research (see "Deliberative Polling"). These various models provide some starting points for further experiment and evaluation; others will certainly be found. Only empirical investigation can determine which of them can best help us to move closer to the democratic ideal.

✍ Deliberative Polling

In the past two decades, there have been several attempts to realize the ideal of deliberative democracy, in which citizens do not simply register their

(cont).

prior opinions, but are provided with opportunities to discuss with one another and to learn from experts before they cast their votes. Deliberative polling, devised and implemented by the Stanford political scientist James Fishkin, is a prominent example, one that has been focused on a variety of topics in different parts of the world.

Deliberative polling brings together people who are selected on the basis of their answers to a prior questionnaire. The goal is to have a varied sample of participants, representing diverse perspectives on the issue to be discussed, and to avoid introducing people who will insist dogmatically on their own original beliefs or who will seek to dominate the conversation. The discussants have opportunities to exchange ideas and arguments with one another, and to listen to the testimony of experts, to whom they can also pose their own questions. A vote on the issue is held at the beginning of the process, and another vote is taken at the end.

One striking example of a change in opinion came in a deliberative poll on U.S. foreign policy. Initially, only a small minority of the participants knew that foreign aid accounts for less than 1 percent of the U.S. budget; at the end, a clear majority had come to recognize this fact, and the original majority favoring reducing foreign aid had been replaced with a majority advocating increasing aid. Deliberative polling could easily be adapted to increase public understanding of the evidence for the scientific consensus in areas that are currently matters of controversy (e.g., climate change). Indeed, it has already been used in Europe to focus on the question of the urgency of tackling climate change.

Deciding What We Know

The last few sections have asked how science should be directed in a democratic society: What does a commitment to democracy tell us about how the agenda for scientific research should be set? A different problem about the relationship between science and democracy can be discerned, though, in some of the fiercest current public debates about science: What does a commitment to science tell us about how our democracy should proceed? We, as a society, accept some beliefs in the sense that we use them as the basis for our decision making about matters of public policy. Who should decide which beliefs we thus collectively endorse? Should we entrust this task to a special caste of scientific experts? Or should all members of the society who wish to make their own judgments on the evidence be able to participate in an open debate, to decide what conclusions we should draw concerning contentious scientific issues?

A long tradition has it that where disagreements are difficult to resolve, free and open debate offers a solution that is both democratic in spirit and epistemologically sound. Allow all views a hearing, this tradition says, and the truth will emerge victorious. But a look at the public disputes about issues such as evolution or climate change quickly dispels this hopeful notion. In disagreements like these, where the disputants are separated by deep differences in their beliefs, values, and standards of evidence, fruitful debate is difficult. Far more common is a free-for-all in which the loudest voices dominate, and onlookers—lacking the knowledge that would give them good grounds for evaluating the outcome—fall back on social trust and intuitive common sense to guide their judgments. These are poor guides in such circumstances, tending to confirm and deepen the onlookers' initial prejudices, or to render them vulnerable to manipulation by those who favor the techniques of persuasion over the presentation of evidence.

Free and open debate does not reliably lead to public acceptance of beliefs that help people to reach their goals and realize their projects. On the contrary, under conditions that are all too common, it tends to undermine these democratic aims. This can be seen clearly in the case of the climate change debate, where a few vocal spokesmen for a marginal scientific viewpoint have helped to foster an illusory controversy, with the result that many people (in the United States, in particular) have come to conclusions that lead them to make choices inimical to their manifest interest in ensuring that the world their children and grandchildren inherit is a good one to live in (see "Climate Change Controversies").

ᔕ Climate Change Controversies

Among climate scientists there is almost complete consensus with respect to the following minimal claim:

(MC) If immediate action is taken to limit the emission of greenhouse gases, the rise in the earth's mean temperature by 2100 will be at least 2°C; if no such action is taken, the likely rise is between 3°C and 6°C.

The evidence for (MC) is based, in part, on measurements of the planet's mean temperature in recent years, and on the use of proxies to calculate the mean temperature for the period before systematic measurements were made. The calculations and measurements support the famous "hockey stick" graph shown in Figure 6.1.

(cont).

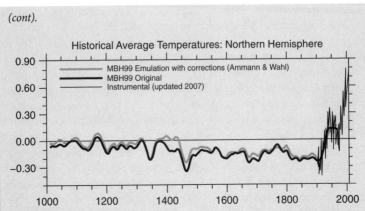

Figure 6.1 *A hockey stick graph. Adapted from C. M. Ammann and E. Wahl,*
2007: The importance of the geophysical context in statistical evaluations of
climate reconstruction procedures. Climate Change, 85, 71–88.

This shows how the temperature has increased since the Industrial Revolution, but it does not by itself specifically support the hypothesis that the warming trend is caused by human activities (that it is *anthropogenic*). However, studies of carbon dioxide concentrations (obtained from tiny air pockets trapped in Antarctic ice) have enabled researchers to recognize the covariation of the concentration of carbon dioxide in the atmosphere with the global mean temperature over a period of 650,000 years. So extensive a correlation is (to understate) highly unlikely to occur by chance.

Yet correlation, even fine-grained correlation, isn't causation. It is reasonable, on the basis of the "hockey stick" and the impressive covariation, to conclude that there is some causal process behind these phenomena, but not to infer that the increased carbon dioxide concentration causes the warming. That last step is made on the basis of two other pieces of evidence. First, there is a known mechanism by means of which higher concentrations of some gases—"greenhouse gases," including carbon dioxide and methane—increase the retention of heat. Second, there are no known alternative causal mechanisms that would account for the correlation: Suggestions of rival causes (e.g., sunspot activity) have been carefully analyzed and refuted. The consensus among climate scientists rests on recognizing a causal basis for the correlation between carbon dioxide concentration and global mean temperature, and seeing that there is only one candidate mechanism, the well-understood trapping of heat when greenhouse gases are emitted into the atmosphere.

Why, then, is there controversy? Part of the answer is that wealthy corporations interested in continuing the widespread use of fossil fuels are prepared to invest both in advertising and in funding scientific research, in order

(cont).

to cast doubt on the reasoning we have reconstructed. They have two main ways of proceeding. One is to contend that the graphs showing changes in the Earth's mean temperature are inaccurate. Recent measurements are hard to argue with, so the focus here is typically on the status of the proxies that are used to investigate historical temperature variations. Although climate scientists have careful arguments for the reliability of the markers they use for calculating past temperatures (e.g., tree-ring data), the considerations that lead them to compute particular values are often not easy to explain to a lay public. Furthermore, the situation has been further complicated by an episode in which e-mails among climate scientists were hacked—those e-mails have been selectively quoted, and the quotations distort their sense, to give the impression that the data have been doctored. A second strategy for climate deniers is to invoke some rival mechanism that accounts for the correlation between carbon dioxide concentration and global mean temperature. However patiently climate scientists show that the alleged alternative process cannot account for the data, there are always possibilities for amending the story, combining it with another alternative, or thinking up something new. These tactics are reminiscent of the campaign against evolutionary theory, where the reward for clearing up one supposedly crucial challenge is to face a new one.

The minimal claim, (MC), is very well supported by the evidence available to us. It is, however, important to understand what (MC) does not imply. Knowledge that the Earth's mean temperature will rise as the result of our addiction to using fossil fuels does not tell us which regions will become warmer, which ice sheets are likely to disappear, how great the rise in sea levels will be, and a host of further things we would like to know. It is reasonable to conclude that many major ice sheets will melt, that the seas will rise, and that low-lying areas of the world's continents and islands will be inundated. It is also clear that extreme weather events are likely to increase in frequency, and that storms, droughts, and heat waves of a severity that used to be rare will become significantly more common. Precise predictions about the future climate of a particular geographical region, by contrast, depend on complex interactions, modeled in different ways by different scientists, and with respect to these it is hard to reach firm conclusions, even about probabilities.

Yet it is worth thinking about what we need to know. The general shape of the threats global warming will bring can be discerned. Future decades are likely to be marked by sudden floods, periods of intense heat and drought, wildfires, disruptions of agriculture, contamination of water supplies, modified patterns of disease transmission, ecological disruptions that allow the evolution of new pathogens, and the periodic destruction of infrastructure in storms we would now count as unusually violent. Nobody has well-grounded advance estimates of the frequencies of any of these scenarios. Nevertheless, it is entirely

(cont).

reasonable to believe that all our descendants, wherever they may happen to live, will be extremely lucky if they manage to avoid most of them.

That pessimistic conclusion should be set against another great uncertainty. Many of those who oppose serious attention to the problem of climate change do so because they fear the cure would be worse than the disease. Draconian measures to limit emissions of greenhouse gases would disrupt the global economy, creating a situation in which future people, although not vulnerable to the climate of a hotter planet, would be drastically impoverished. In comparison with attempts to model the climate and weather patterns of particular regions in coming decades, economic predictions about the future course of the world economy are even more speculative. There is, nonetheless, a serious worry about economic development in a world that is thoroughly committed to limiting greenhouse gas emissions. The issue is further complicated by the need to take into account the needs of poorer nations, which are seeking to improve their situation by industrializing, and to recognize the demands of global justice. The real climate change question is not whether (MC) is true, but how we balance the uncertainty of the climatic threats we face against the uncertainty of the economic future, subjecting both to the constraint of dealing justly with all members of our species. That question tangles together issues in natural science and social science with questions in methodology, ethics, and social philosophy.

A deeper examination of the climate change debate illustrates how well-ordered deliberation can take account of the prominent place of values in some disagreements about science. People with radically different views about climate change often agree that the debate has been distorted through the improper importation of values. Gripped by the ideal of science as value free, they chastise their opponents for distorting the evidence in light of the goals those opponents want to promote. Because the value-free ideal is a myth, this common presupposition must go—but that still allows a critique of the improper importation of values. To see this, consider two versions of the debate. In the first, a passionate environmentalist, devoted above all to the preservation of some particular species, insists that immediate policies for curbing the emission of greenhouse gases are needed, whereas a researcher whose salary is paid by Gargantuan Oil is equally adamant that there is nothing to worry about. In the second, both those who urge policies for limiting emissions and those who oppose them share a concern for the future well-being of people on our planet: The one side stresses the rising sea levels, the increasing frequency of extreme weather events, the disruptions of agriculture, the prevalence of drought, and the likelihood of pandemics; the other side emphasizes the collapse of the global economy, the consequent social upheavals, and so forth.

Values enter (on both sides) into both debates. The difference lies in the fact that the values involved in the first debate are idiosyncratic, formed on the basis of personal ties and short-term interests, whereas the values that figure in the second are attentive to the most central concerns of people around the world: to wit, the predicament that they and their descendants will confront. Both scientists in the first debate would be legitimate targets of criticism. The second—which should start from the reality of global warming caused by human beings—is the sort of debate we should be having.

A free-for-all among uninformed people with strong passions is unlikely to help with difficult issues. On the other hand, turning judgments over to an elite group of experts is problematic because the experts' goals and values are unlikely to be representative of those of the public at large, and may lead those experts to conclusions that—once again—fail to serve the public good. Even where this problem does not arise, public perception that scientific elites are motivated by values that are not widely shared rules out automatic reliance on experts, if the central ideals of democracy are to be honored.

The solution must be a middle road between these alternatives—making science itself more inclusive of diverse viewpoints and more responsive to the goals and values of the public, yet respecting its distinctive authority. We learned from the feminist critique of science that even as individuals we can take responsibility for our epistemic situation, and make choices aimed at broadening its perspective. But a more important lesson concerns the social organization of science. If there is no coherent hope of achieving a comprehensive "view from nowhere," our science should aim to achieve a view from everywhere—one that incorporates the insights and aspirations of a wide range of perspectives, in active interplay with one another. This means, at a minimum, drawing a wider range of people into the ambit of scientific work and decision making than we have done so far and giving serious consideration to how their interactions should be structured to take advantage of the diverse perspectives that they contribute.

At the same time, effective two-way communication with the larger public is essential for good judgment within the sciences and good uptake of scientific judgments in the public arena. One important medium of communication can be conceived along the lines already sketched—groups of people representing a wide range of viewpoints who are led "behind the scenes" so that they can see clearly the kinds of judgments that scientists are making and the reasons for them, and who are called on to press their own questions and concerns so that scientists can understand and respond to them. Other channels of communication are important as well. Both science journalism and science education can contribute far more effectively than they now do to the cultivation of public scientific literacy and to overcoming the misconceptions fostered by the value-free ideal.

Conclusion

We ended the introductory chapter of this book with the image of science as a human activity. In this final chapter we have developed an image of that kind to illuminate how issues of values in science can best be addressed in democratic societies in which market institutions are powerful and science is influential in education, government, and business. We have tried to open a perspective on how the core values of other, earlier images of science can be carried forward. The sciences still need to make sense of the world, to explore newly opened frontiers of understanding, and, where, possible, to accumulate more accurate and detailed explanations of complex features of the natural world. Values are central to all these efforts. Issues about how to fit scientific research into democratic societies—how to integrate the knowledge of experts with ideas about democratic freedoms—are unresolved. We close this book with a clear recognition that, on this, as on many other questions, we have only begun a discussion. We hope readers will not only be moved to continue thinking about the questions we have raised—and often left unanswered—but will also be inspired to think philosophically about the practices of the sciences and the relation between science and other social institutions, posing questions the philosophical tradition has not yet confronted.

What we call science emerged from the activities of small groups of well-to-do seventeenth-century gentlemen, dedicated to pursuing their own curiosity. Neither they, nor most of the researchers who came after them, could have foreseen the ways in which science would develop, how it would change, how it would come to conflict with many traditional beliefs, and how it would become central to the functioning of complex democratic societies. Philosophy of science, as we understand it, should aspire to the broadest and deepest reflections on this institution whose practices, standards, and self-image—hugely consequential as they are—are all products of happenstance and serendipity. If philosophy consists in critique, it is critique designed to amend and improve. We hope to have illustrated some of the ways in which such critique can be conducted.

Suggestions for Further Reading

A groundbreaking discussion of many of the issues discussed in this chapter is Helen Longino, *Science as Social Knowledge* (Princeton, NJ: Princeton University Press, 1990); Longino's views are further articulated in *The Fate of Knowledge* (Princeton, NJ: Princeton University Press, 2002).

The proposal that the sciences aim at significant truth is further developed in Philip Kitcher, *Science, Truth, and Democracy* (New York: Oxford

University Press, 2001). Heather Douglas, *Science, Policy, and the Value-Free Ideal* (Pittsburgh: University of Pittsburgh Press, 2009) is now the *locus classicus* for discussions of the ideal of science as value free. Hugh Lacey, *Is Science Value Free?: Values and Scientific Understanding* (London: Routledge, 1999) is another illuminating treatment of the place of values in science, with special emphasis on the issue of control.

Vannevar Bush's profoundly influential report, *Science—The Endless Frontier*, is now available in an edition with a valuable introduction by the historian Daniel Kevles (Washington, DC: National Science Foundation, 1990). Maurice Finocchiaro provides a clear—and riveting—view of Galileo's conflict with the papacy in *The Galileo Affair* (Berkeley: University of California Press, 1989). David Joravsky, *The Lysenko Affair* (Cambridge, MA: Harvard University Press, 1970) is lucid and thorough. In *The Great Betrayal: Fraud in Science* (Orlando, FL: Harcourt, 2004), Horace Freeland Judson provides a detailed survey and analysis of examples of scientific misconduct. The notion of well-ordered science is introduced in Kitcher, *Science, Truth, and Democracy* and later elaborated and defended in his *Science in a Democratic Society* (Amherst, NY: Prometheus Books, 2011).

James Fishkin explains and motivates deliberative polling in *When the People Speak* (New York: Oxford University Press, 2009). A brief overview of the approach, and some reports of results is available at http://cdd.stanford .edu/polls/docs/summary/. A useful overview of participatory research in the context of environmental management is Dianne E. Rocheleau, "Participatory Research and the Race to Save the Planet: Questions, Critique and Lessons from the Field," *Agriculture and Human Values*, 11, 1994, 4–25.

For discussions of climate change and the issues it raises, see (besides the sources recommended for Chapter 1) Stephen Schneider et al., *Climate Change: Science and Policy* (Washington, DC: Island Press, 2010).

Index